LORRAINE WEISS

d/MW[b]	d	RI[c]	BP	UV cutoff[d] nm	Cost[f] £/l	Notes
0.50	1.70	1.267	~57.0	210		Fluorinert FC78 (3M), Flutec PP1 (ISC) mainly perfluoro-n-hexane, expensive.
0.81	1.517	1.250	47.6	230	1.5	Arklone P (ICI) stabilizer free.
0.87	0.629	1.358	36.2	210	2.5	low bp, more expensive but less toxic than hexane, miscible with CH_3CN in all proportions.
0.61	0.692	1.404	99.3	210	2.0	Less aromatics, higher bp and more viscous than hexane, more miscible.
0.80	0.659	1.375	86.2	200		Good UV.
0.93	0.779	1.427	81.4	210	2.2	Rather viscous but different selectivity from hexane.
1.07	0.740	1.406	49.3	210	20.0	Too expensive at present.
0.93	0.640	1.371	30.0	215		Not readily available.
1.66	1.260	1.626	46.2	380		Too smelly and toxic, not UV.
1.04	1.590	1.466	76.8	265		Rather viscous, poor UV.
0.83		1.413		225		
~0.82	0.864	~1.500	139.1	290		Not UV.
0.71	0.747	1.368	89.64	220		Peroxide hazard, rather low boiling.
1.09	0.862	1.378	35.74	225	10.0	Too expensive, butyl chloride similar.
0.94		1.496	110.6	285		Not UV.
1.00	0.873	1.397	68.3	225	8.0	Expensive.
0.98	1.106	1.525	131.7			Not UV.
1.13	0.879	1.501	80.1	280		Not UV.
1.31	1.460	1.424	38.35			Too expensive.
0.71	0.724	1.368	68.3			Peroxide problems.
0.96	0.713	1.353	34.6	220		Peroxide hazard, too low boiling.
0.93	0.836	1.442	92.1	290		Not UV.
1.26	1.500	1.443	61.2	245	2.0	UV not ideal.
1.57	1.336	1.424	40.1	235	1.8	UV not ideal.
1.23	0.880	1.408	66.0	215	3.5	Peroxide problems.
1.27	1.250	1.445	83.0	225	2.2	Best UV of chlorinated solvents.
1.12	0.805	1.381	79.6	330		Not UV.
1.12	1.001	1.400	131.2	380		Not UV.
1.36	0.818	1.359	56.5	330	1.0	Not UV but good solvent and cheap.
1.17	1.033	1.422	101.3	220	6.3	Purification needed to achieve good UV.
1.02	0.901	1.370	77.15	260		UV not ideal.
1.25	0.927	1.362	57.1	260		UV not ideal.
0.92	0.815	1.410	137.8	210		
1.40	1.100	1.478	189.0			Much too viscous except as a minor component.
1.10	1.022	1.586	184.4			
0.97	0.702	1.387	55.5	275	2.0	Limited UV, but basic.
0.72	0.728	1.401	89.5			Basic.
1.85	1.138	1.394	101.2	380		Too expensive, not UV.
1.91	0.782	1.344	82.0	190	5.0	Expensive but otherwise ideal.
1.24	0.983	1.510	115.3	305		Very limited UV, but basic.
0.95	0.863	1.380	83.0	220		
1.32	1.306	1.552	180.48			Used hot (typically 130°C) for permeation chromatography.
1.11	1.460	1.476	87.19			Permeation chromatography (PC)
1.30	0.949	1.428	153.0			Permeation chromatography.
0.98	1.620	1.505	121.0			Permeation chromatography.
	0.819	1.470	202.0			Permeation chromatography.
0.72	1.025	1.618	235.0			Permeation chromatography at high temperature (165°C), less corrosive than OBCB, stabilized by topanol OC.
1.07	0.937	1.438	165.0			
0.648	0.896	1.470	194.6			
1.39	1.390	1.219	73.6			Permeation chromatography.
~0.77	0.965	1.401	124.6	220	2.0	Good UV if pure enough.
1.34	0.786	1.380	82.3	210	1.8	More viscous than ethanol and methanol, but popular additive.
1.71	0.789	1.361	78.5	210	0.3	Excise duty problems; more viscous than methanol but more miscible.
2.49	0.796	1.329	64.7	205	1.0	Cheap and excellent UV if pure enough.
1.80	1.114	1.427	197.3	210		Much too viscous.
1.75	1.049	1.372	117.9			Useful additive—acid.
1.80	1.000	1.330	100.0			

d Wavelength at which transmission falls to 10% (1A) for good commercial LC grades, glass distilled at prices shown.

e Values from L.R. Snyder[8], with additions from Riddick and Younger[20], and Barton[128] generally at 20°C sometimes by interpolation.

f Approximate cost in small containers at suitable purity for LC. Prices vary widely with quantity and purity.

LIQUID CHROMATOGRAPHY

in Practice

P A Bristow

ICI (Pharmaceuticals Division)
Macclesfield, UK

hetp

Published by **hetp**, 10 Langley Drive, Handforth, Wilmslow, Cheshire, UK, SK9 3BQ
hetp is registered in England No 2001926

Printed by H Oldfield & Son Ltd, Macclesfield, UK

Typeset in Univers on an IBM Electronic Composer
By Janet Hutson of Print Co-ordination, Macclesfield

Photographic work by Heywood Photographic, Hazel Grove, Stockport

ISBN 0 9504833 0 3 paperback
0 9504833 1 1 hardback
0 9504833 2 X microfiche

UDC 544.92

"Therefore, since brevity is the soul of wit, and tediousness the limbs and outward flourishes, I will be brief:—"[29]

PREFACE

Thanks

My main thanks must go to my employers, ICI (and that has meant the personal support of many individuals who make up the organisation) for giving me the chance to find out about LC over the past six years. Their library facilities have been invaluable. I thank ICI for permission to publish this book, and especially for allowing me to use **hetp** as my publisher.

Another large contribution has been made by all my co-practitioners, very many of them in ICI, who have contributed by their discoveries, corroborations, difficulties, failures and questions.

Information for this book has come from too many sources to list them all, apart from the very important contribution made by the formal literature. The following people deserve particular thanks for giving information or reading the draft text and finding many errors: F Bailey, P N Brittain, J V Dawkins, R A C Gray, M J Holdoway, J J Kirkland, J H Knox, A Mackay, J C Meakin, A Pryde, M G Rayner, A Titterton, E S Watson, B F Williamson.

Some illustrations are reproduced by kind permission of the copyright holders: the authors and publishers of the Journal of Chromatography, Journal of Chromatographic Science and Analytical Chemistry. This debt is acknowledged implicitly by the reference to source with each illustration.

Many thanks must go to those involved in physically producing the book: D Davies, L Oldfield and R Heywood patiently taught an amateur much about printing. Janet Hutson transformed my mis-typed text into beautiful copy which was a continual encouragement to continue writing. Patrick Bill checked some of the calculations. My parents (non-chromatographers) gallantly proof-read much of the book. I have often read authors' thanks for their wives' forbearance and thought they were exaggerating: they were not — and I thank mine for hers.

Objectives

This book tries to present, in a palatable form, the know-how needed by *practitioners* of the art. Know-how we still need to know will often be discussed, for the state of the art is, and should be, still far from static even after nearly a decade of rapid development. Conciseness is for the reader's sake, and reflects the author's prejudices, but enforces omissions that a review ought to include. References to papers and reviews will allow the reader to fill these gaps.

It is not intended to supplant the more detailed, rigorous and authoritative texts, most notably of Kirkland and Snyder[3], but to find a place on the bench of the chromatographer and provide a compact guide to the practice of LC.

The aim is to explain general principles and yet to be specific about 'how to do it'. The presentation is designed to support this objective. Rigor is sacrificed for simplicity: theory for practice.

summary **• each summary is an over-generalisation**

The summaries are chosen to help the reader judge *whether* to read a section, as an aide memoire *after* reading it, or even as a *substitute* for reading it. The summary aims to say all one *needs* to know.

The problem of depth is one of the most serious facing an author. I have tried to consider each facet of LC at several levels to allow for the varying degrees of interest and expertness. The most detailed level reflects my interest and expertness. A still deeper level lies with the expert users of LC: sometimes they describe their work in the literature. Personal communication is also sometimes possible, and usually helpful.

A determined attempt has been made to present all theory and equations in a form suitable for 'back-of-chromatogram' calculations, especially by including units, the lack of which is a powerful deterrent to putting finger to calculator.

Examples have been chosen sparingly. Many have been shown too often before and so new examples have been sought. The vast LC literature is full of examples: the references given are one point of entry. Perusing the current journals will reveal the latest, and hopefully the best and most relevant examples.

This book is not, however, about specific applications: the rapidly expanding literature, and reviews and other books are available to meet this need. Applications will always tend to be very specific to individual users and the need to know how to develop a new separation will always remain.

CONTENTS

1 INTRODUCTION

Liquid Chromatography is chromatography with liquids. This trivial definition has merit in emphasising the unity of mechanisms of all methods of separation induced or eluted by liquids as the mobile phase. Column, thin-layer, paper, ion-exchange, gel permeation, and the more fashionable forms described variously as high pressure, high speed, high efficiency—or more cynically, high price—liquid chromatography are all practical and useful exploitations of a single physical process described by a single theory. For separation the so-called stationary phase must have a different ability to hold (adsorb) different compounds against removal (elution) by the liquid: its bulk form, paper, silica, or resin, solid or liquid, differs for each variant but its precise nature where separation really happens is often much less certain.

This book is really about high resolution liquid chromatography (LC) and how to use it in practice, but will compare and contrast with these other ways of using the chromatographic process with liquids, especially when they have lessons for LC. There is a danger that separation expertise acquired in old LC will be lost in the enthusiasm for the new and then much effort wasted on rediscovery in the future.

The Chromatographic Process

The chromatographic process is fundamentally the same in all the many ways it is applied. Thin-layer chromatography(TLC) is a visible process: what goes on in a column is very similar, but hidden by the column tube, usually stainless steel, and the packing material itself (unless the eluent happens to have the same refractive index as the packing). Processes visible near the wall in glass columns tend to be atypical of the rest of the column area.

No diagram can match spotting a TLC plate with a few dyes and a mixture and watching it run.

As the capillary action 'pumps' the liquid up the plate, an applied single component sample spot moves up the plate. As it moves, the zone becomes broader; the further it moves the broader it gets. Although different compounds move at different speeds, they are all broadened to about the same extent.

The mixture spot first moves up as a whole, then elongates, and finally starts to separate. The further the 'solvent front' moves, the more the spots spread out and separate. Fortunately they separate faster than they broaden. If the solvent flow is stopped by lifting the plate above the reservoir liquid (equivalent to turning off the pump in a column system) the spots stop moving, and do not broaden or spread out perceptibly even if the plate remains wet by being kept in the solvent-saturated environment of the tank.

If the plate is dried and scanned with a densitometer, a chromatogram will be obtained. However this chromatogram is not quite the same as one which will be obtained from the detector at the end of a column.

The detector in column chromatography cannot, alas, see the whole plate at once but only senses the solutes which *leave* the column. It is like viewing a TLC plate covered by an opaque sheet with a narrow slit: as the plate is run, the spots may move beneath the slit and be detected.

A disadvantage of column chromatography is clear: there is no information about components that have yet to leave the column. This is counter-balanced by the high sensitivity, linearity and range of LC detectors, and that much more liquid can be passed than is possible on a plate because the solvent front reaches the top of the plate (unless multiple development techniques are used).

Basic features of a Liquid Chromatograph

There are five basic components of a chromatograph:

- pumping system
- injection system
- column — separating system
- detecting (and recording) system
- collection system

[The omnibus term 'system' is used to include the many possible devices which vary widely in complexity, capability and cost.]

These items are discussed in detail in section 7, the possible materials and separation mechanisms in section 6, and how to pack them in section 4.

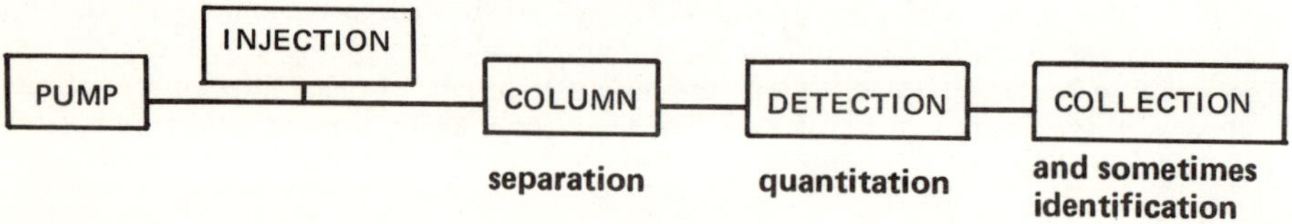

Operation

The pump forces the eluent (solvent or mobile phase) over the so-called 'stationary phase', often an adsorbent like silica. The pump may be capillary action on a TLC plate or paper chromatogram, gravity for a classical column, or one of many types of pump for HPLC.

The sample (solute molecules) to be analysed is injected onto the top (start) of the column (or plate) and eluted by the eluent.

The column should separate individual solute molecules (components) because they are each retained to a different extent. Thus they are eluted at different speeds along the column and emerge from the column at different times.

Components which emerge in solution from the bed can be detected and their elution time and quantity assessed and recorded. (Only with an open bed 'column' like TLC can all components be detected, in principle at least, during elution.)

The eluting liquid can be collected and solutes recovered by removing the solvent. This is usually described as preparative chromatography, even though the column may simultaneously be used for analysis.

A Simple Liquid Chromatograph

A simple instrument (on which research of the highest quality can be performed, despite the fact that it might not be described in the catalog as a 'research model') might consist of:

- a direct gas-driven coil pump
- a syringe injection device fitted to
- a short column (100-200 mm) packed with 5 or 10 μm diameter silica and
- linked via 1/16 in diameter tube to a
- UV detector, for simplicity a fixed wavelength of 254 nm type, but equally simply for selectivity a variable-wavelength model
- samples eluting might be collected from the detector outlet in small vials for subsequent identification
  (Hardware is discussed in detail in section 7)

A typical chromatogram using an instrument of this type is shown. The sample is a rather impure technical grade of 3-isopropylphenol. It contains many alkyl substituted phenols in addition to the main compound and other reaction products including various quinones. This chromatogram shows the power of LC.

- resolving power – about 20 peaks can be separated
  (several are off-scale on this chromatogram)
- speed – the complete analysis takes only a few minutes
- sensitivity – a total of 200 μg was injected; many peaks represent only nanograms of impurities
- range – the major peak would be several metres high
- quantitation – peak areas can be measured accurately
- identification – spectra of peaks can identify components (for example see section 7)
- preparation – each separated component can be collected

(see below)

Developing an analytical method

- choose a suitable column configuration and operating conditions
- decide on a separation strategy (see section 8.1)
  and choose a suitable separating material (see section 6)
- pack a suitable column (section 4 suggests methods)
- choose an eluent (mixture) (see sections 8.2 and 12.5).

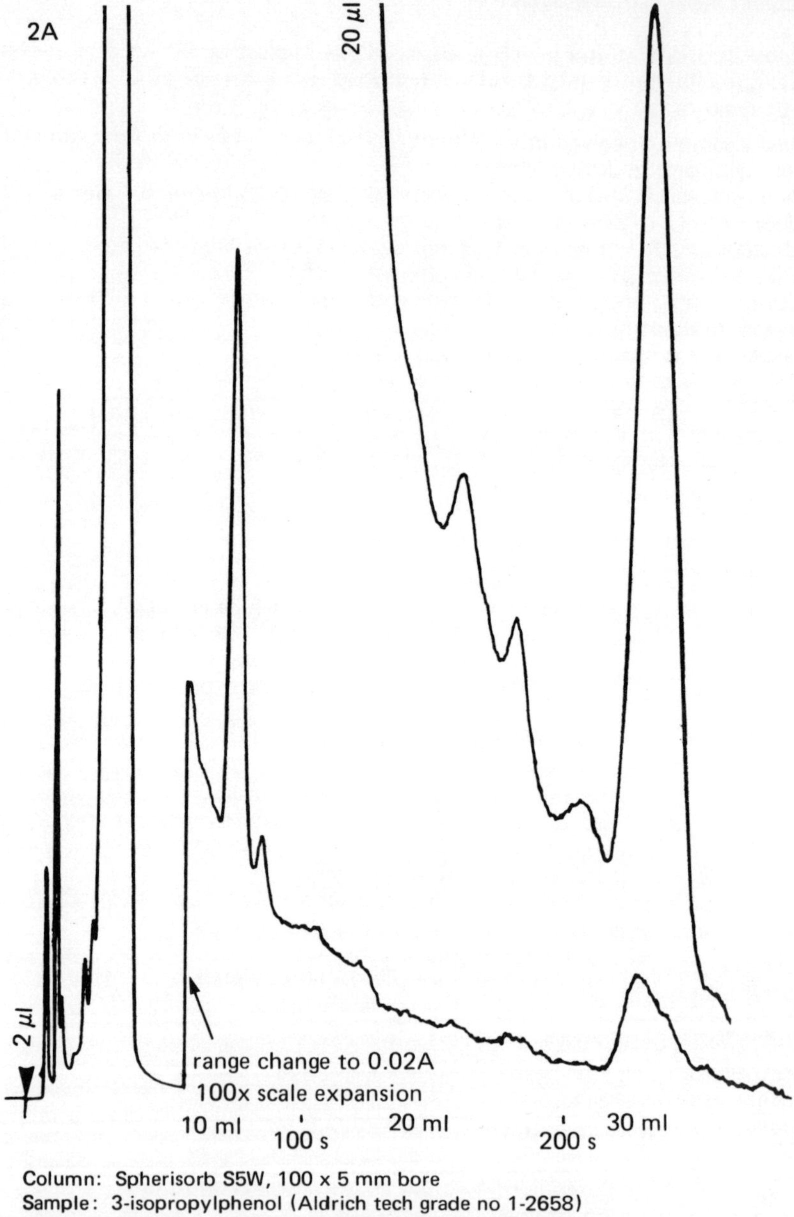

Column: Spherisorb S5W, 100 x 5 mm bore
Sample: 3-isopropylphenol (Aldrich tech grade no 1-2658)
Eluent: hexane:ethanol (99:1 v/v) (Rathburn Chemicals Ltd)
Pressure: 71 bar Flowrate 0.156 ml/s Chart Speed 0.5 mm/s
Detection: Cecil Instruments CE212, 200 nm 2A and 0.02A
Pump: CE210 coil pump, nitrogen gas driven

LC in practice P A Bristow **hetp**, Macclesfield UK ISBN 0 9504833 0 3

Operating the chromatograph

- fill the solvent reservoir, perhaps after degassing solvents
- start pumping liquid until a stable baseline is established
- inject a test mixture to check the system is working correctly
- inject a sample dissolved in the eluent, say 10 μl, perhaps with flow stopped
- calculate from the flowrate and diameter (see sections 3 and 11.5) when one 'column volume' will have passed and mark this on the chart. (Differential or solvent peaks may appear near this point. Any peaks eluting before this point are excluded from the pores of the material.)
- continue eluting until 10-20 times this volume (time) has passed.
  - ▲ if no peaks appear, the solvent system is not strong enough
  - ▲ if all peaks appear bunched together near the eluent time marker, then the eluent is too strong (too polar for a polar adsorbent)
  - ▲ if all peaks tail see section 12.1 – or for any other faults
- alter solvent composition if necessary
  - ▲ if changes in the eluent do not increase the selectivity of separation, try changing the packing material (simply changing to another manufacturer's silica is unlikely to make sufficient difference to be worthwhile).

When a satisfactory chromatogram is obtained:

- analyse the chromatogram by measuring peak area or height (see sections 8.3 and 7.7)
- establish response factors by injecting solutions of known concentration of authentic materials
- calibrate using internal or external standards (see section 8.4).

If the precision or accuracy, speed, solvent consumption or pressure needs are not satisfactory:

- alter the detector conditions (range and wavelength) to increase, decrease, or equalise sensitivity – try to get peaks to the same size when measuring their ratio
- alter the detector pathlength
- use a different *principle* for detection (see section 7.6 or 8.10)
- alter the injection volume (see section 8.7)
- change the column operating conditions, or size if necessary to save time, solvent consumption per analysis, or to reduce pressure requirements
- consider column switching techniques (section 8.9) if there are very many analyses to be done.

To identify unknown peaks:

- identify with the detector, for example a *scanning* UV detector
- collect peaks emerging from the detector and use MS, IR, NMR (even analytical columns containing a few grams of packing material can separate a milligram or so)
- use a (second) highly selective detector, for example atomic adsorption or flame emission for N and P
  - ▲ more than one technique is usually needed
- confirm by spiking with an authentic sample.

Selecting column parameters (suggested starting values in brackets)

length (100 mm)
- increase for more plates, say up to 1 m, if separation selectivity proves difficult to improve by changing separation system (packing material and solvent)
- increase to increase the weight of packing material when collection of samples is the aim
- reduce length for very fast (but easy) separations, for example to 50 mm
- reduce to cut solvent consumption if the degree of separation permits and there are very many samples to be analysed
- avoid coupling columns

bore (~5 mm for microparticles)
- increase to ~8 mm, or up to ~25 mm for prep
- increase to ~8 mm to ease injection or detection volume problems
- decrease to reduce solvent consumption (provided the detector can cope with smaller peak volumes)
- decrease if absolute sample size is limited, say to 1-2 mm, though this may require a smaller detector cell volume
- decrease if the cost of the packing is very high, often only a consideration for very long columns and pellicular packings
- if the flowrate that the pump can deliver is limiting, it may be better to reduce the diameter if $\nu < 1$

particle size (5 μm)
- increase to reduce cost for preparative work
- increase to allow a longer column to be used if pressure capability is limiting, perhaps to increase the weight of packing material for prep
- use 30–50 μm pelliculars only if a separation system cannot be found with silica or bonded phases on microparticles

pressure
- vary only to attain a suitable flowrate (subject to constraints of the pump and injection technique)

flowrate
- adjust, perhaps by varying pressure, to give $\nu = 5$ (see section 3)
- increase speed as much as convenient and feasible considering resolution losses (and limits on detectors response speed)
- avoid decreasing ν below 1

column switching and back-flushing
- consider for complex mixtures, particularly if they contain components of widely different properties
- consider for reducing analysis time
- consider back-flushing for separating whole classes of compounds
- avoid unless there are many analyses to be carried out. Developing a method may not prove straightforward and will need some specialised engineering

Note that it is rarely worthwhile changing any parameter by much less than a factor of two. For example doubling the length or halving the particle diameter may double the plate number but this only means a 50% increase in resolving power.

Choosing a solvent for the sample

summary
- ● always safest to dissolve sample in the eluent
- ▲ and to use a large injection volume if necessary
- ● the polar component of an eluent mixture may give satisfactory results
- ▲ but may also lead to anomalies and poorer separation

The solubility in the eluent (or often the *low rate of dissolving*) of the sample to be analysed is a practical problem. The temptation is to use a much more polar (stronger) solvent to dissolve the sample. This often has a deleterious effect on the separation, as illustrated below:

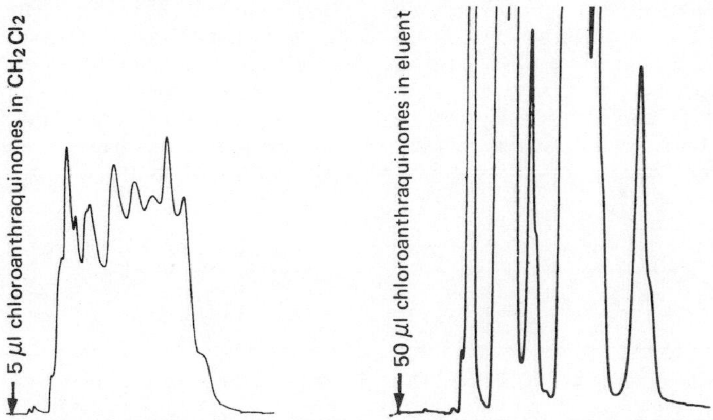

Eluent: 0.5% CH_3CN in hexane
Column: Spherisorb S5W, 100 mm long, 5 mm bore

The injection of 5 μl of dichloromethane (a large load if viewed as a solute injected on to a small analytical column) produces a wave where the equilibrium adsorption of the polar constituent of the eluent (in this case acetonitrile) is disturbed. The profile of the acetonitrile concentration in the column effluent as the dichloromethane elutes might be:

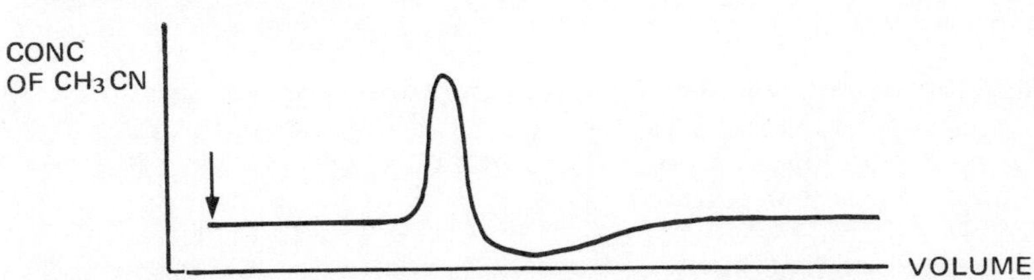

The effect of this dis-equilibration is difficult to predict but is rarely desirable, if only because the separation will change with injected volume.

The second chromatogram above shows that it is better to dissolve the sample in the eluent and inject a much larger volume (ten times in this example) of a more dilute sample.

LC in practice P A Bristow **hetp**, Macclesfield UK ISBN 0 9504833 0 3

'Solvent' peaks

summary
- **unexpected peaks near solvent front may be artifacts**
- **changing eluent composition also causes artifacts**

An unexpected feature of many chromatograms is the presence of peaks which do not appear to correspond to any solute. The peaks are often negative-going, or have derivative form. They are often dismissed as 'solvent' peaks: a little more explanation is usually possible.

Experiments by Slais and Krejci[224] show that injected solutes cause local changes in the eluent composition which are sensed by the detector. They used a two-component eluent of cyclohexane–diethyl ether (98:2 v/v), injected various test solutes and monitored with a RI detector. All solutes produced a peak, either positive or negative, with capacity factor $k' = 1.2$. Diethyl ether has the same capacity factor in this sytem. A solute like nitrobenzene that is retained *more* than than ether caused a peak showing an increase in ether concentration, whereas a solute like benzene which is *less* retained caused a negative concentration peak. The size of the peak was correlated with the weight of solute injected.

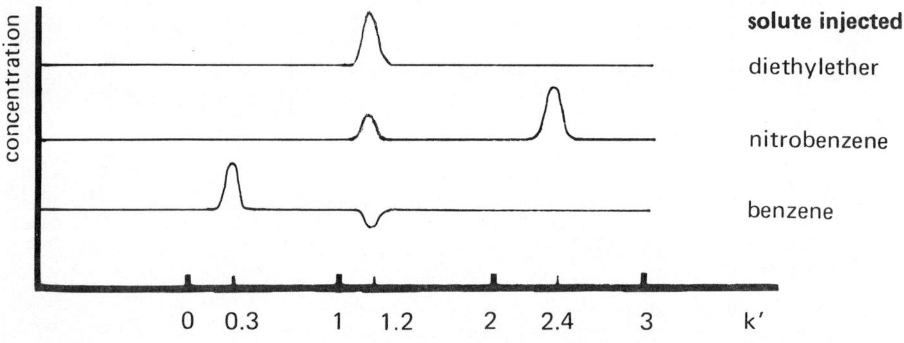

In practice the detector response is often more complicated. RI detectors respond positively and negatively (and the actual chromatograms shown in reference 224 are more complex than the diagram shown above). UV detectors respond not only to absorbance, but also to refractive index changes, producing differential peaks.

If the 'solvent' or 'vacant' peak interferes with the chromatogram, Slais and Krejci recommend changing either to a weaker polar additive so that the vacant peak comes first, near to the solvent front, or to a much stronger one so that it occurs much later (and broader) than peaks of interest.

Injecting the individual components of the solvent system, and the solvent used to dissolve the sample, is a useful check for artifacts. (It is wise to make absolutely sure that the syringe is clean.)

With stepwise and gradient elution even more complicated artifacts are common. Engelhardt and Elgass[141] warn of some of the pitfalls and urge a test run without solutes in the injected liquid. Solvent impurities can become unexpected solutes.

Reducing retention on a polar adsorbent by making the eluent more polar

Eluent: ethanol:hexane (concentration of ethanol shown)

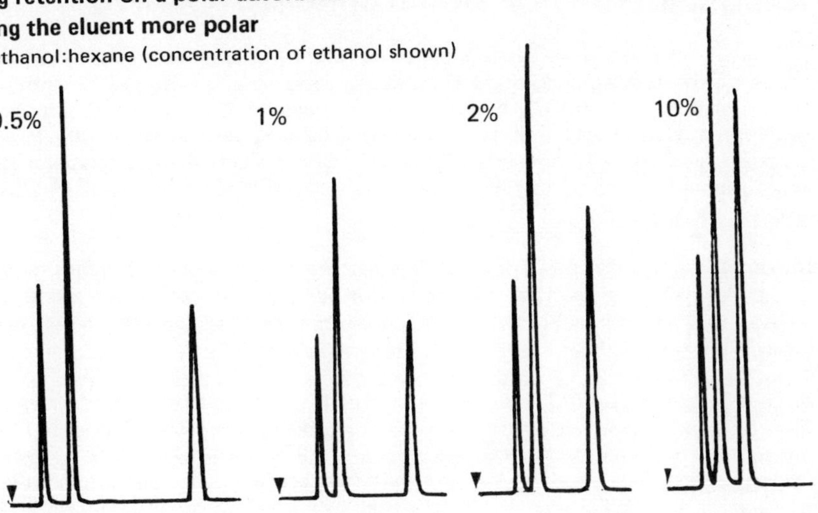

Predicting how much polar component to use

Based on Scott & Kucera[375]

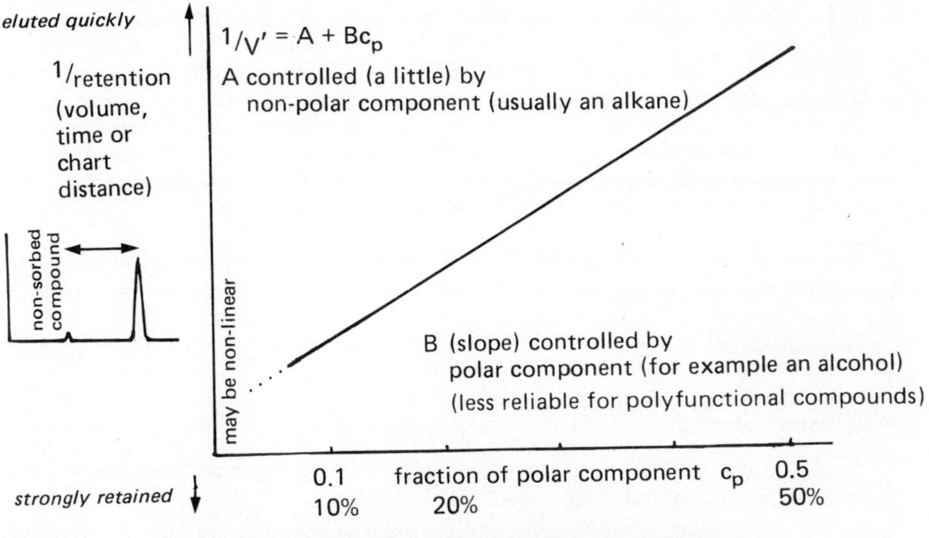

summary

- **doubling polar constituent of an eluent very roughly halves retention of well retained peaks on silica**
- **small concentrations of polar components may have a dramatic effect compared to none of that component**
- **changing the non-polar part, usually an alkane, has a slight effect on retention and selectivity, but may alter miscibility ranges**

LC in practice P A Bristow **hetp**, Macclesfield UK ISBN 0 9504833 0 3

2 THE HISTORY OF LIQUID CHROMATOGRAPHY

It is really too early to write about the history of chromatography though there are some interesting accounts of the story so far[11,165]. It certainly promises to be fertile ground for future historians of science especially those pondering why developments took place at certain times–and not before. Hopefully to indulge in a little speculation is not too premature. One objective in such speculation is to avoid making the same mistakes again: rediscovering facts is wasteful; rediscovering errors is inexcusable.

Chromatography has had some dormant periods: was it lack of technology, theory, or lack of men of action and conviction to provide push, and money, or even fashion that let it lie idle? Why did the development of gas chromatography surge away leaving LC to catch up only in the last decade?

A first dormant period followed the discovery of chromatography in 1906 by Tswett[166], a Russian botanist who used column chromatography to separate plant products, for example carotin and xanthophylls, eluting with petroleum from a calcium carbonate bed. A very similar separation was almost re-invented by E. Lederer in 1930. Even the next decade saw a development only slightly faster than stagnant.

Between 1940 and 1950 the fundamental theory and most fundamental, if primitive, experimental implementations of liquid and gas chromatography were thought out and proven, if not refined. The name of Martin stands out: his contribution was recognised by the award of a Nobel Prize in 1952. Gas chromatography advanced steadily and then spread very fast to nearly every laboratory as soon as the flame ionisation detector was both invented and taken up by instrument manufacturers. By contrast LC made little real advance except in amino acid analysis when Moore and Stein produced their most successful reaction plus photometry detector[167], and in TLC where detection was easy because zones were either directly visible, or could be developed in some way.

During this period LC was accepted as low resolution, large particle, large column, low speed and low pressure – if with resignation by Martin and Synge[40] in 1941

" . . . in striving for conditions for uniformity of flow, the high pressure and small particles desirable for smaller HETP have to be abandoned."

even though the theory clearly said

" . . . the smallest HETP should be obtainable by using very small particles and high pressure difference across the column."

In the fifties when GC was advancing rapidly it is difficult to see why theory was ignored. Technology cannot be the sole limitation, for in retrospect one can check that all the items needed, even in todays instruments, could have been produced at that time, though now many components are more readily available or better. Pumps were made and *could* have been made in corrosion resistant steels or ceramics; stainless steel compression fittings were less readily available but are not essential; some methods of producing fine, well sized adsorbents were known and others required research, not borrowed technology. Detectors alone emerge as deficient and needing instrument development.

Detector development needed different skills from those of most chemists who were the users of LC. Exceptional biochemists with crucial analyses like Moore and Stein might solve their own problems but still not make the technique widely used. Only instrument manufacturers could develop equipment for all to use. Many companies produced gas chromatographs and spectrometers, and several even produced both, but their managements failed to grasp the nettle and invest in research in LC in the early sixties when all the necessary technology and expertise was available.

Many of the pionèers saw that the photometric detector was the only one likely to prove sensitive and robust enough for general use. But they were not supported by photometer designers: their failure to meet the need for small aperture, long pathlength flowcells in their optical path geometry imposed a considerable practical barrier to using conventional spectrometers as LC detectors. The use of optical null systems rather than continuous readout was another problem.

This made design of suitable flowcells difficult and thus prevented the use of smaller particles and columns.

Yet their decision must receive some sympathy: at that time it must have seemed unlikely that many would buy spectrometers as detectors when peaks were emerging so slowly that one person could collect samples from several columns and measure them by hand using one spectrometer with conventional cells. But without detectors, columns with small particle size packings could not be developed.

So two workers whose detectors were to have a wide impact as commercial versions turned their backs on the spectrometer approach and designed their own systems. Kirkland turned to the process analyser of Dupont which had filters selecting lines from a mercury lamp. It was quite simple to adapt this for micro-flowcells and it had very high sensitivity and stability. From this prototype came a complete chromatograph, the Dupont 820, and others followed. Watson, having failed to convince his management that LC should be a development area[168], produced a design as an independent consultant which was to become the basis, under licence[78,96], of a whole generation of commercial 254 nm instruments from many companies. Though in some ways primitive, this design was simple and cheap to make and allowed the use of 8 μl flowcells. It was originally designed for use with nucleic acid analysers meeting the needs of the Horvath–Lipsky pellicular resins[42] but differed from the many other similar detectors then available in having a suitable long pathlength micro-flowcell.

Later two detectors based on spectrometers which were fully variable wavelength appeared. One was based on the Zeiss instrument and came from the use of this spectrometer by Huber. A cheaper instrument was based on the Cecil Instruments spectrometer: this was designed to be used with small aperture cells, though even its designer, C S C Tarbet, had not envisaged quite as small cells as were installed. These two European instruments were never widely sold in the USA and led to a widespread use of wavelengths other than 254 and 280 nm over this period: this is reflected in the published literature until the mid-seventies when variable wavelength detectors became widely available in the USA.

Other detectors being developed during the late sixties made much less impact on the development of column and separation technology mainly because their sensitivity was too low. Thus refractometric detectors were invaluable for permeation chromatography but their promotion for adsorption chromatography counter-productive for the columns then available because their large particle size and low efficiency, or small capacity, demanded a much higher sensitivity.

The development of pellicular supports was a useful advance for chromatographic separations but exacerbated the detector problem. Their low capacity meant that long columns were needed and injected sample sizes could only be small, so the concentration of solute reaching the detector was very low. The pellicular concept is only a way of obtaining a surface a few micrometers thick. Much effort by many groups and companies went into perfecting this type; unfortunately it was at the expense of development of particles which were only a few micrometers in diameter—which would not fill most of the column with inert, but zone spreading, glass spheres. Historically, pellicular materials lie at the end of an evolutionary branch, destined to become extinct.

In retrospect (always an easy vantage point) the problems in preparing and packing are not as difficult as they seemed at the time. Two pre-conceptions from GC needed to be shed. One was that particles could be sized by sieving; the other that dry packing was appropriate for microparticles. Interparticle forces are proportional to surface area, which might be over a hundredfold greater than for typical GC packings. These forces must be neutralized by dispersal with a liquid into a slurry to permit either sizing or packing. Neither sedimentation sizing nor slurry packing was novel at that time.

So the slow progress of high performance LC seems to be due to a combination of factors particularly the interlinked requirements of the column packing material and packing method, and most important, the need for a suitable detector. The pioneers of LC had to solve both of these problems together in order to produce systems that would compete with conventional column chromatography, thin-layer chromatography, which naturally works with slurry packed microparticles, and GC. The pioneers of GC by contrast had none of this competition and achieved dramatic successes very early on.

It is also tempting to speculate that many people whose talents might have been used to develop LC were preoccupied with developing GC: different skills and attitudes from the users of traditional column chromatography were clearly necessary to solve these problems simultaneously.

Permeation chromatography has developed on several surprisingly separate and parallel paths from other branches of LC[23]. One path was the much needed molecular size distribution analysis of polymers; another the separation of much larger biological molecules on very different gels, from which also came the tailored selectivity of affinity chromatography. J Porath and P Florin[263] proved the aqueous separation process which has been invaluable for the separation of very large molecules, but the gel materials were not suited to non-aqueous solvents and higher temperatures essential for dissolving synthetic polymers.

J C Moore saw the need for rapid size fractionation of the many polymers being developed at that time and set out to prepare suitable gels. First successes were recorded in 1962 and a classic paper[262] presented more detail in 1964. Waters Associates licensed these cross-linked polystyrene materials from Moore and the Dow Chemical Co and added a vital ingredient, an effective commercial chromatograph, and the technique took off. These patents, and aggressive marketing, gave them a tight grip on the market—one which history may eventually decide was too tight to be best for science. As with other fields of high technology, the symbiosis of science and commerce is essential, but it has its price.

Looking to the future (which is more dangerous than looking back) it is difficult to see any fundamental ways of improving the technique. Likely improvements are more in details affecting reliability and robustness than in pushing back the limit of what is possible. Knox and Saleem[248] looking ahead in 1969 noted that *"While pressures of 3 000 psi can readily be generated in liquids, formidable technological problems will have to be solved before columns 10 cm long containing 2 micron diameter particles can be operated without loss of performance."* These problems are almost solved and we are now near to many of the limits[140].

3 THEORY OF LIQUID CHROMATOGRAPHY

Theory is boring: but useful in practice. The slow development of LC is a cautionary tale about ignoring theory. This book aims to provide only the theory that a practising chromatographer needs to know. Many other texts and papers present more thorough, rigorous and erudite treatments of the interesting but incomplete theory of chromatography. The treatise of J C Giddings[1] must be singled out as a summary of many important advances on the early but nevertheless definitive basic theory of Martin and Synge[40]. L R Snyder's[8] book represents a similar advance in recording our understanding of the separating surface in chromatography, though some more recent contributions, many by Snyder himself, are needed to complete a picture of the current state of the art. J H Knox has updated the application of theory; his book[4] 'Applications of High Speed Liquid Chromatography' contains a brief but lucid summary of the non-equilibrium theory of chromatography.

The nub of the problem of developing a complete theory is that chromatography is essentially a near but still non-equilibrium process. The flow is (and must be to avoid significant zone spreading by simple diffusion) too fast to allow equilibration of the solute molecules between the eluent and the stationary zone.

Even the problem of flow of liquid through a bed of inert particles is difficult to model well because it depends on how the bed is packed. Most chromatographic results, the main source of data, are an average of the whole bed and tell little about the detailed flow between particles and into the pores of the particles.

In combination these microscopic flow patterns and molecular diffusion factors pose problems which still challenge theoreticians. Nevertheless small elements of theory guide research for practical improvements. For better chromatography the aim must be to improve equilibration rates to get nearer to equilibrium. Increasing molecular diffusion speeds and reducing distances molecules must diffuse is one area where improvements are sought, and found. Improvement in the flow patterns, through better methods of packing the particles into the column, is another.

summary

- **mixing and dilution in column of inert particles exemplifies flow pattern problems of injection, packing, fittinas and plumbing, and detection**
- **equilibration between packing surface and eluent determines separation**

An over-simplified view is to divide the problem into two and ignore any interaction as a practical approximation. (It also over-simplifies the experimental problem of finding inert particles, especially porous ones.) A column of inert particles (possibly porous) exhibits the *diffusion and flow pattern* problems of chromatography but will not separate at all. *Separation* is achieved by the chromatographically active surface layer. The surface interacts with the solvent, and may include solvent molecules (and may be thick enough to penetrate not only into the pores, but into the body of some types of particle). Separation is determined by the equilibration of sample (solute) molecules between the stationary surface and mobile eluent: the theory of separation is discussed further in section 5.

Solving a problem with chromatography requires both separation by choice of packing material and eluent composition; and engineering eluent flow through this material so that the separated components are not diluted or mixed up again too much.

In time it should be possible for the user to place himself in the hands of the suppliers with confidence that column hardware and packing material will all give performance as good as the best that can be achieved. Alas the rate of technological change in both areas has been, and will be for a few more years, too fast for the user to ignore the need for performance evaluation. To make valid and fair comparisons some understanding of the theory is essential: careful control and measurement of experimental variables is also necessary.

Most of the practical problems in applying chromatography lie in the flow pattern area: with injection, column packing, fittings and plumbing, and detection. Most of the skill in chromatography lies in selection of the best packing material and the best eluent mixture to solve a problem. It is usually easier to achieve separation by chromatographic skill, but unnecessary to make the problem more difficult by introducing inefficient flow patterns from negligent engineering.

The 'Inert Particle' Column

An ideal column of inert particles would allow one to inject a sample slug at the top and for it to emerge undiluted with unaltered profile at the column end. (And ideally the detector should respond and record this profile without distortion.) Real columns broaden and dilute the sample zone. Glass beads are usually the experimenters first model for inert particles. An important aim of column design is to reduce this zone broadening because it dilutes the sample, reducing sensitivity and tends to mix up again any components that have been chromatographically separated into distinct bands.

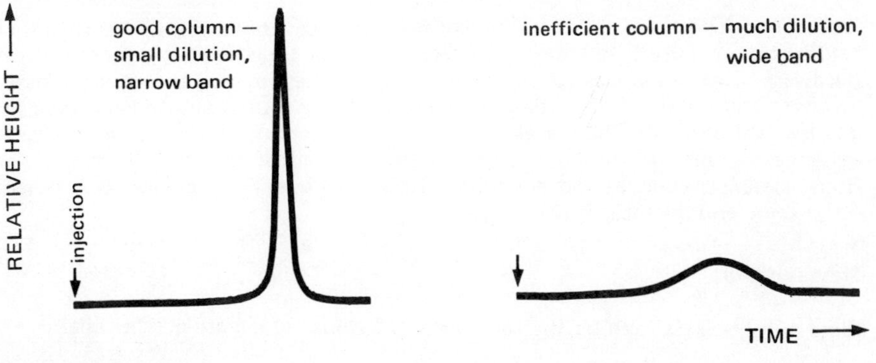

Zone broadening

Theories of chromatography assume square sided injection slugs which are predicted to be broadened to Poisson and then Gaussian distributions. The concepts of plate number and plate height are left over from analogous theories of distillation columns. Plate number remains as a confusing but universally adopted measure of the amount of zone broadening (dilution and mixing) a column causes.

Standard Deviation – Zone Width

summary • **standard deviation ≈ 1/4 of zone width**

Assuming the zone is gaussian (rarely strictly true in practical LC), the diagram below shows that zone width may be taken in practice as 4 standard deviations (4σ).

The standard deviation (σ) and the variance (σ^2) can be measured in time (σ_t), distance (σ_z or σ_L), volume (σ_v or σ) or even in practice, chart distance, and this is a cause of some confusion. (Subscripts will be used to refer to specific units but otherwise volume units can be assumed.) The plate number concept applies regardless of these units (since it is dimensionless) and compares the peak-width with distance, volume, time, length, time, or chart distance from the point of injection. (Since compressibility of the liquid can be ignored, a non-linearity complicating the concept in GC is absent.) The standard deviation σ can also be visualised as the peak 'quarter-width'. The link with statistics is only a mathematical one and is irrelevant to chromatographers.

Variance (σ^2)

summary • **σ is peak quarter-width–may be in various units**
• **variances σ^2 are additive**

Theoretical interest in variances is that they are additive

$$\sigma^2_{\text{overall}} = \sigma^2_{\text{column}} + \sigma^2_{\text{injection}} + \sigma^2_{\text{detector}} + \ldots\ldots\ldots$$

That standard deviations, and thus variances, can be expressed in so many different units has diminished practical use: unfamiliar units of volume variances (m^6) are an added disincentive.

Volume variances are most useful because they can be easily derived from chart traces and measured flowrates, and there are clear links with the geometry of the hardware. Time variances can be more suitable for estimating the effect of detector response speeds or a finite injection time. Since we should aim to keep the column *volume* variance as the weakest link, the injection, detection and plumbing variances should be minimal, and at most a third of the overall peak variance. Some detailed calculations are given elsewhere: injection, section 8.7; detection 7.6; tubing and detector flowcells 12.4.

Plate number

Plate number is defined by the following equations, which are all equivalent

$$N = L^2/\sigma_L^2 \quad = tr/\sigma_t^2 \quad = V^2/\sigma_v^2 \quad = Z^2/\sigma_z^2$$

Assessment of plate number is simple from a detector chart recording. Several methods have been used[17], but the quickest one giving least scope for artistic licence is measurement of the chart distance from injection and 'peak width at half-height'. The diagram below shows most of the relationships.

Plate number efficiency in chromatography

$C = 1/\sqrt{2\pi}\, e^{-\frac{1}{2}(x^2/\sigma^2)}$

where $x = t - t_r$

σ = standard deviation

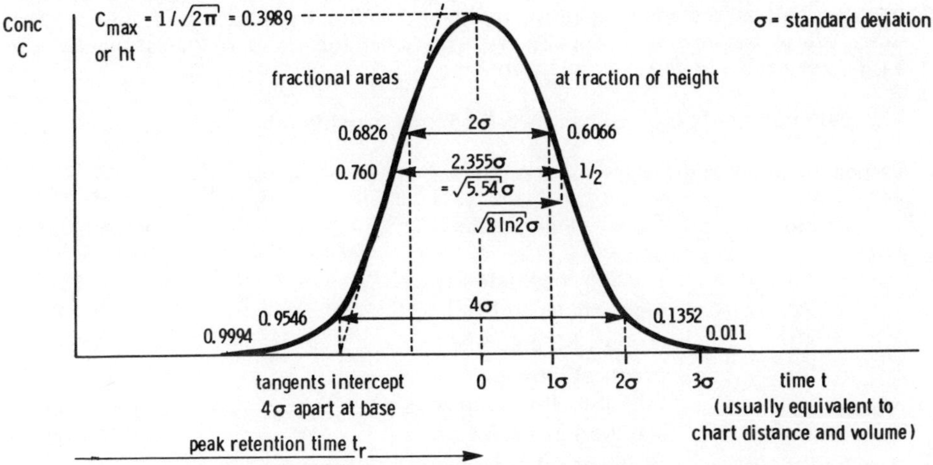

Plate number $N = (t_r/\sigma)^2 = 16(t_r/4\sigma)^2 = 5.54 \times$ (or $2^2 \times 2\ln 2$) $\left(\dfrac{\text{retention distance}}{\text{width at } \frac{1}{2} \text{ ht}}\right)^2$

Height-width Values for Gaussian Peaks

fraction of height	distance from peak*	peak width at height*
.99	.01	.02
.9	.46	.92
.607	1	2
.5	1.175	2.35
.2	1.79	3.58
.135	2	4
.1	2.15	4.3
.05	2.45	4.9
.01	3.0	6
.001	4	8

*in units of σ

Note that these figures can be used to check how well the peak shape fits the gaussian function. Thus the ratio width at half ht/width at tenth ht = 2.15/1.175 = 1.83

Dispersion Ratio

The ratio of the distance from injection to peak maximum/half-height width may be described as a dispersion ratio. In that it is directly related to resolution it is more useful but less theoretically tractable than the standard deviation. Section 11.8 gives a table of ratio and plates numbers.

The plate number N is given by $N = 5.54 \times (\text{dispersion ratio})^2$

Values for some columns, real and predicted, are listed below.

Ratio	Description	Plate Number
10	starting chromatography	554
20	high efficiency LC	2 000
40	current LC columns	9 000
100	practical limit ? 500 mm with 3 μm packing	55 000
135	achieved in 1975	100 000
4000	Giddings calculated max plate number	$\sim 10^8$

The plate number introduces a squaring which confuses in practice: doubling the plate number only increases resolution by 50%.

Plate number is especially useful (and popular) in that it can be determined without knowing anything about the column dimensions or even flowrate. Most other parameters require additional data. Erroneous results can be deduced from chromatograms if the eluent composition changes during the chromatogram, or if there is a significant lag between the column end and the detector, or flowrate changes.

summary • **plate number measure of column performance**

Plate Height

summary • **plate height H is smaller for**
- **better columns**
- **small particle diameter**
- **less viscous solvents**
- **higher temperatures (but separation may also be less)**
- **small sample molecules**
- **slow flow velocity (but not too slow)**

Plate height, or 'height equivalent to a theoretical plate', HETP or H, follows naturally from the plate number concept.

$$H = L/N = \sigma_L^2/L$$

This distance compares the amount a zone has spread with the distance the zone has travelled. Obviously it should be as small as possible and has become a prime measure of the column efficiency (or, more realistically, inefficiency).

Remembering that the base width is defined as 4 standard deviations, the distance over which a peak is dispersed at the end of a column of length L is $4\sqrt{HL}$. So short (small L) and efficient (small H) columns spread and dilute zones least.

The relationship $N = L/H$ links the early measured plate number N directly with length and plate height, and shows that equally well packed columns with twice the length should have twice the plate number. This does not always turn out to be true in practice, that is H increases and the column is said to be less efficient.

Because the plate height is primarily a function of the particle size, the so-called reduced plate height is a useful parameter which corrects for this factor.

Reduced Plate Height

Reduced parameters were first proposed by Giddings[1, 41] and have been popularized by Knox[4]. These apparently abstruse parameters (that only theoreticians could invent) are simple and practical to use.

The reduced plate height $h = H/d_p$. It is the number of particles per plate. Regardless of particle size it is about 2 for well packed columns. In practice reduced plate height can be determined from

$$h = H/d_p = \text{column length (mm)}/(\text{no of thousand plates} \times d_p\ (\mu\text{m}))$$

Thus a column 200 mm long packed with 10 μm particles and having a plate count of ten thousand would have reduced plate height of 2. Like other reduced parameters, h is dimensionless.

Zone Broadening

The way in which the microscopic flow patterns spread out the zones is complex and little understood in detail though many aspects are intuitively fairly obvious. The column consists of many flow passages outside and inside the particles. The passages may have widely differing shape and size, and consequent resistance to flow. Flow will be faster in wider sections than in narrow gaps, and like a river, individual flows will swirl about and change with time. Clearly it is difficult to apply theories such as those developed for long tubes of circular cross-section to these assemblies of irregular channels. One can only be sure that flowrate will be strongly dependent, perhaps to the fourth power, on diameter.

These flow effects outside the particles are classified as eddy diffusion. The structure of the packing determines the effect of eddy diffusion on actual column performance and this is why so much practical work has revolved around the shape, size and packing of the particles. Velocity inequalities which lead to zone spreading have been attributed to a variety of modes:

- trans-channel variations occur because the flow near the wall of each channel is small compared to the centre flowrate;

- short-range interchannel effects result because a packed bed consists of tightly packed regions several particles across and larger channels about a particle diameter across bridged by jammed-in particles, and high flowrates will be found in these large channels;
- longer range interchannel effects are caused by uneven distribution of these tight and loose packed zones;
- trans-column flowrates over the whole column will vary because of wall effects, where looser packing and higher velocities are usual, or because the column tube is bent to form a coil (rare in LC).

All of these modes interact so the result, and any theory, must be complex. The several interesting theoretical treatments have borne little practical fruit.

Molecular Diffusion into the Pores

Flow through the pores is negligible because their diameter is so small compared to the spaces between the particles which are similar in size to the particles themselves. Access to the pores is only via molecular diffusion. So coupled with the relatively large scale eddy diffusion is molecular diffusion across and between channels, and in and out of any microscopic pores into particles. To give some idea of scale, most particles are tens of *micrometres* in diameter, and most pores tens of *nanometres*, a thousand times smaller. The shape, length and diameter of the pores is obviously important but this is an area of which we know too little.

Reduced Velocity

The reduced velocity $\nu = u\, d_p/D_m$ takes account of the particle size d_p and liquid diffusivity D_m. It compares the flow velocity down the column u with the speed of molecular diffusion into the pores of the particle D_m/d_p.

When the two velocities are about equal the column operates most efficiently. This relationship will be discussed further in detail but one can see that if flow speed is too high the equilibration by diffusion to and from the pores and their surface will not be able to keep up.

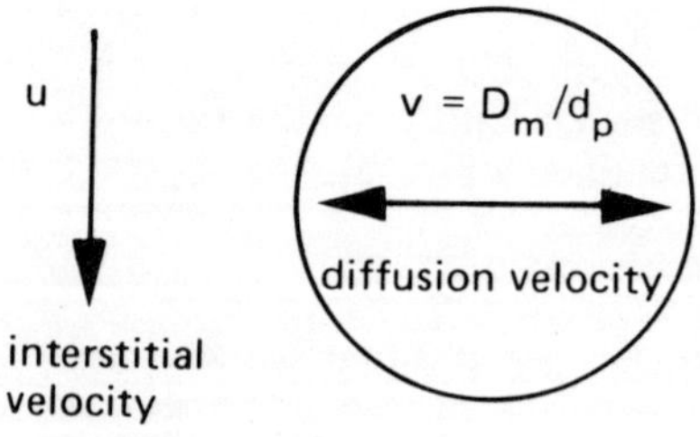

reduced velocity $\nu = u/v$

Inserting some typical values for LC: $d_p = 10\ \mu m = 10^{-5}$ m and $D_m = 10^{-9}\ m^2/s$

$$v = D_m/d_p = 10^{-9}/10^{-5} = 10^{-4}\ m/s$$

which is comparable to an interstitial velocity u of 0.1 mm/s. Thus under these conditions a molecule is about as likely to be moved by diffusion as by flow.

Diffusion Rates

summary • **low eluent viscosity and small solute molecules diffuse faster**

The diffusivity of a solute in the mobile phase D_m is a function of the solute as well as the solvent liquid. In theory one needs to know this parameter for each solute in each solvent, but in practice the measurement is much too tedious and most workers rely on a few dubious extrapolations from the few experimental determinations in the literature. It is clear that large molecules diffuse more slowly and hence the molar volume appears in many equations for calculating diffusion coefficients. The most popular is the Wilke-Chang equation[45] (though citation seems more popular than calculation). Some calculations are presented in section 12.6. Fortunately in practice the effect of even quite large errors in the diffusivity on the application of theory is quite small. Viscosity of the eluent is clearly most important in determining the speed at which solute molecules diffuse. Hydrogen bonding has a marked effect—Wilke and Chang introduce a 'correction' factor of 2 for water.

The practical significance of diffusivity is usually small, as other factors controlling the equilibration rate are usually much more important. As a rule one should nevertheless select a solvent with as low a viscosity as possible: this also helps to reduce the pump pressure. It may also help in difficult cases to increase the temperature which decreases viscosity and hence increases diffusivity.

Thus the drive is towards smaller and smaller particles, for the actual distance travelled is less and thus the analysis time is also reduced. The trend to smaller particles has been halted by the limits on the size and speed of the injection and detection hardware and it is difficult to see how these can be improved dramatically. The main reduction of particle diameters a hundredfold from 200 μm to 2 μm has taxed hardware severely: it is difficult to visualise a practical realisation of even the next tenfold reduction in particle size though materials of this colloidal size have produced separation[46].

It may help to summarise the dimensional factors of a column packed with 5 μm porous microparticles. If the diameter is 5 mm the column is a thousand 5 μm particles across, and a layer consists of $\sim 10^6$ particles; 40 000 particles can be placed end to end along a 100 mm length; and it contains $\sim 3 \times 10^{10}$ particles in all. (Even the most expensive packing materials are quite cheap per particle.) The channels between the particles are a micrometre or so across, but typical pores will be 5 to 10 nm in diameter.

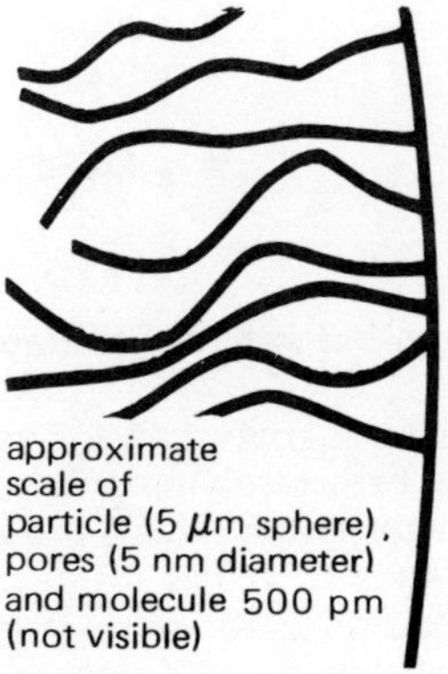

approximate scale of particle (5 μm sphere), pores (5 nm diameter) and molecule 500 pm (not visible)

Reduced Length

A further interesting, but less useful, reduced parameter is the reduced length L/d_p. This parameter expresses the number of particles that can be placed in a row end to end down the column. Each particle effectively does the same amount of chromatographic work regardless of size. So columns which are the same number of particles long should be compared.

Thus a 100 mm column with 5 μm packing has a reduced length of 100/0.005 = 20 000 and should give comparable performance to a column 400 mm long 20 μm packing column (when L = 400/0.02 = 20 000). Each sample molecule passes 20 000 particles from injection to detector. A column diameter equal to one particle diameter is quite impractical but helps display the situation.

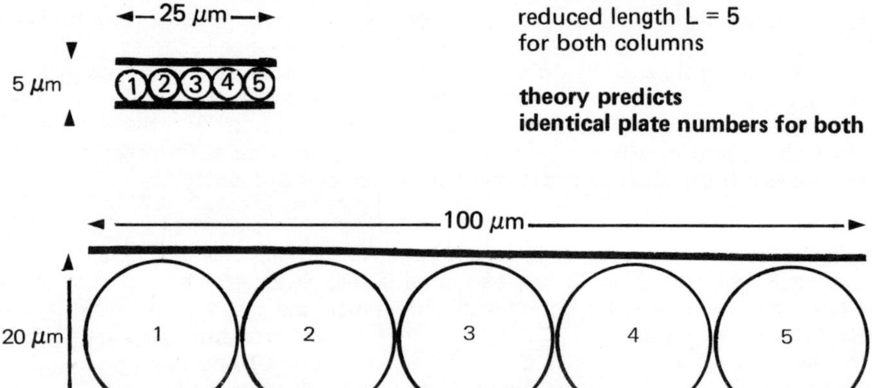

summary

- **reduced parameters** — **dimensionless and corrected for particle size**
- **reduced plate height** — **no of particles per plate**
- **reduced velocity** — **speed down column versus speed of diffusion in and out of pores**
- **reduced length** — **no of particles sample passes from injection to detection**

Flow Velocity versus Plate Height Efficiency

Practical analysts want analytical answers quickly. (They also want high sensitivity and high resolving power—and with as little effort, as applied pressure, as possible). Martin and Synge[40] predicted from their theory that small particles were the key to success. It has taken three decades to turn theory into practice: and now speed is too great for some detectors and their pen recorders. As with GC, detecting and recording speeds impose the real practical speed limits.

Much practical and theoretical work has revolved around finding and modelling the relationship of plate with flow velocity. This was because most work was done with rather large particles so that the theoretically optimum flowrates, when flow and diffusion are at similar rates, were impractically slow. Thus there was a real impetus to speed up the process while sacrificing least efficiency.

With smaller particles, nearer the optimum size, the detector speed limits have imposed a better all-round compromise giving most efficiency at a convenient flow velocity. LC (as GC has done since the early stages) now works naturally near the peak efficiency—and LC turns out to be just as fast as GC and in some ways faster.

There are a number of ways of describing the flow velocity—plate height relationship. A recent paper by proponents of three versions showed their high degree of equivalence—and an agreement to differ amicably by the respective authors[123]. The reduced parameter approach given here is chosen mainly for its simplicity, but its theoretical validity and the quality of supporting experimental data also support its selection.

The general relationship of reduced plate height and reduced velocity is shown by the curve below, which also summarises the controlling factors which differ over the various reduced velocity ranges. This type of plot is now called a Knox plot after its main proponent, Professor J H Knox. The key advantage of reduced parameters is that the form and parameters of this curve are not primarily a function of particle size.

Knox plot **Speed versus Efficiency**

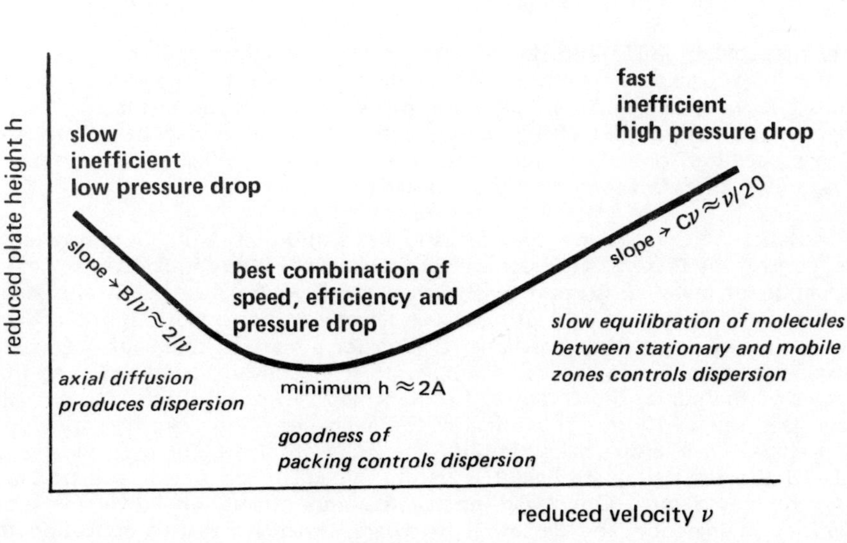

LC in practice P A Bristow hetp, Macclesfield UK ISBN 0 9504833 0 3

The curve can be fitted to the expression

$$h = B/\nu + A\nu^{0.33} + C\nu$$

From experimental data A, B and C can be determined and are useful parameters describing the material and how it is packed. Correlation of values with theoretical models is also possible[4]. Briefly these three factors characterise the zone dispersion from three different causes.

A reflects the dispersion from the spaces between particles—that is how well the column is packed.

B is a fairly fixed and calculable factor (from particle shape and porosity) that shows the zone spreading that takes place by diffusion.

C reflects the dispersion caused by the separation process of adsorption and desorption. If the rate of mass transfer between stationary and mobile states is slow, C is large and at high ν values will dominate, causing a large plate height. This is effectively the dispersion that takes place in the stationary zone inside the pores of the particles.

Obviously, provided the hardware can cope (which it usually can down to about 1 μm), one should go fast enough to avoid a significant contribution from the B/ν term. (Note though that this equation is clearly invalid when B is zero, that is the flow has stopped. Very little zone spreading takes place even overnight with flow stopped so there is ample time to obtain spectral information with flow off, a distinct advantage over GC where 10^4 greater diffusivity and the high compressibility lead to rapid zone spreading.)

Quality of packing determines the A term. It may be helpful to link the A term with the inert particle column concept: under suitable test conditions many columns behave rather like inert particles. Since packing is under the chromatographers control, assessment of this component is important in practice to compare packing procedures, or column purchases. Poor packing is reflected in large plate heights at low speeds but even more at high speeds.

Characteristics of the packing material, and its relationship with the eluent and solute, control the C term. The user will probably rely on published data to guide his selection of material. Current target values for C are ~0.1 or below and most reputable materials achieve this. But because selectivity is so important in practical resolution it may pay to accept a larger C term for a material that separates well: ion-exchange resins are a typical example. If it is convenient to work at the minimum of the curve, the C-term is less important. In practice this means small particle sizes, say 1–10 μm. Thus users of ion exchange resins were first to adopt small particles[105]. Microparticles simplify the users problems, for over the range ν = 1–10 the reduced plate height is hardly dependent on speeds and packing quality the key factor. Only wall or transcolumn effects, and differences or deficiencies in injection and detection hardware, provide a real complication to prevent easy and meaningful comparisons.

 ISBN 0 9504833 0 3

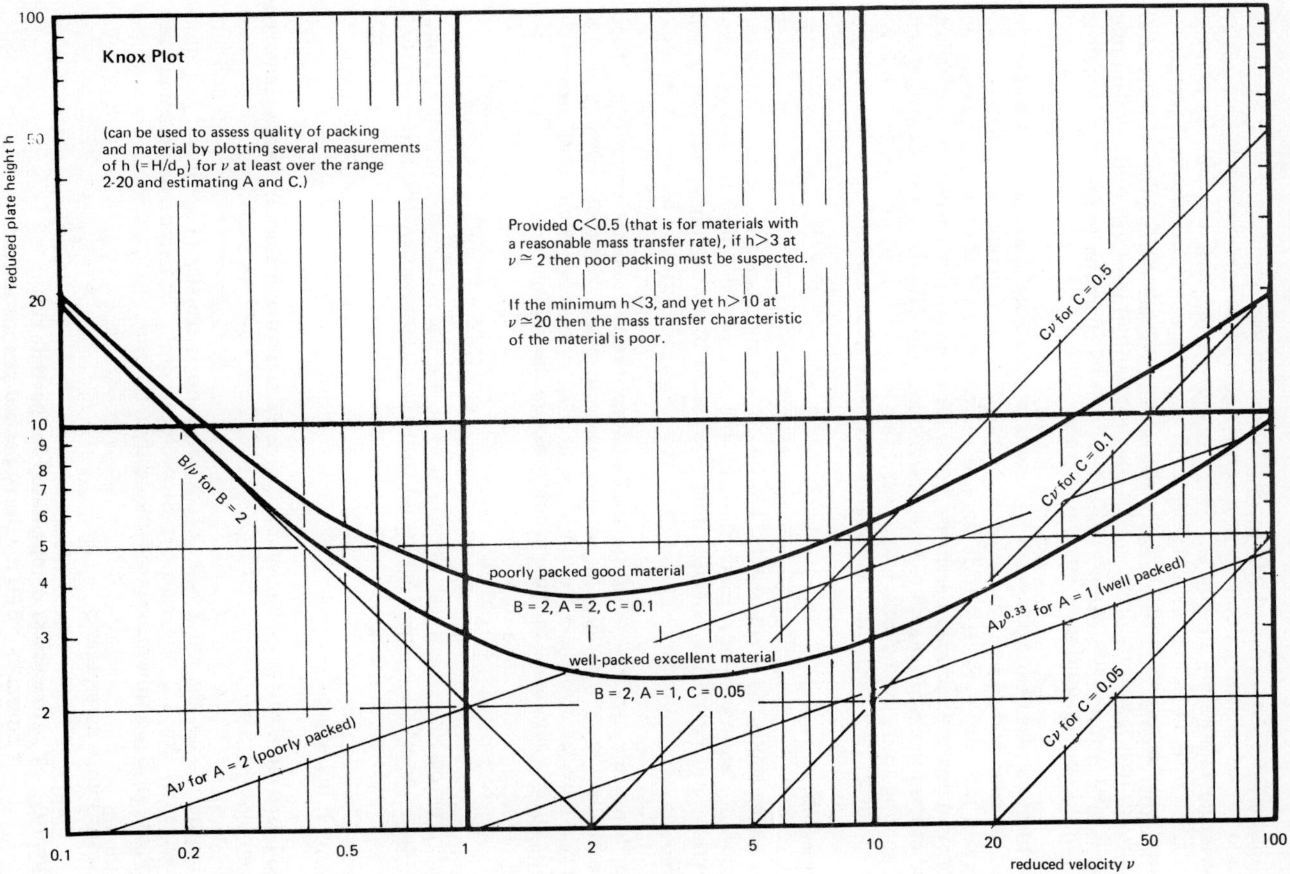

A full packing material evaluation calls for determination of the h – ν curve over a wider range of flow velocities, preferably under ideal conditions where the above interfering factors are minimised. This usually means using the 'infinite diameter effect'[106] thus avoiding wall effects with small central on-column injections in rather wide columns. This data should be comparable with other studies made under similarly ideal conditions. Alas much published data can only be compared in a limited way because it does not meet these criteria. In actual use for analysis a different column configuration may be more suitable.

The formula plate height $H = Du^n$ where u is the velocity and D and n parameters has also been widely used. It has some simplicity but numerical evaluation of D and n is still not very easy. It also represents reasonably some theoretical models, and approximates the reduced parameter equation given in this treatment, but fails to predict the increase in plate height at low reduced velocity. Now that microparticles are coming to be used almost exclusively, this is a real area of interest (which it was not for larger particles) and this equation now seems less useful.

Permeability

The permeability of a column (or conversely its impedance to flow) is important because it determines how much pressure the pump must deliver to achieve a given flowrate, and hence the analysis time – a factor of considerable practical importance.

The flowrate F_c is determined by the pressure (drop) P, eluent viscosity η, length L, and *specific permeability* K° of the column. These parameters are related by Darcy's Law, described by Giddings[1] as the *"Ohm's law of fluid flow"*, where the fluid flux q is analogous to the current, pressure the voltage,

$$q = K^\circ P/(\eta L) = F_c/(\tfrac{1}{4} \pi d_c^2)$$

In chromatography the linear velocity u of an average molecule of eluent is usually of more interest. It is therefore useful to derive a *chromatographic* permeability K by

$$K = u \eta L/P$$

These two parameters differ by the porosity ϵ of the material, usually between 0.4 and 0.8.

Permeability has the dimension of an area and is usually given in m^2, or more conveniently for LC materials the SI multiple μm^2. The (chromatographic) permeability of 5 μm spherical silica is typically 0.04 μm^2.

Flow resistance parameter ϕ'

summary
- **permeability compensated for particle size**
- **normally ~500 for slurry-packed porous spheres**
- **~1000 for non-porous beads, and pelliculars**

As with plate height, velocity and length, it is convenient to compensate for particle size in order to be able to compare the packing density and permeability of columns packed with different sized particles. The dimensionless flow resistance parameter ϕ' is defined by

$$\phi' = d_p^2/K^o = P\,t_o\,d_p^2/(\eta\,L^2)$$

[Some workers have used the inverse parameter K^o/d_p^2 (= $1/\phi'$) which might be regarded as a 'reduced permeability'. However the normal value of ϕ' of ~500 is easily remembered and seems most popular.]

Effective spherical diameter

The practical value of ϕ' is in comparing the packing density of different techniques, for example in comparing dry and slurry-packing like Knox and co-workers[204], or in assuming a value, usually 500, and calculating an 'effective spherical diameter' from $d_p = \sqrt{\phi'\,\eta\,L^2/(Pt_o)}$.

It is this parameter that matters to the chromatographer, not the nominal size marked on the bottle.

Separation Impedance

The concept of Separation Impedance (H^2/K or $h^2\,\phi'$) expresses the separating efficiency of the column for the amount of pressure required to produce it. Low values correspond to high efficiency for a small pressure. Since this parameter is numerically the same regardless of whether it is expressed in reduced terms ($h^2\phi$) or in terms of actual measurements (H^2/K), it avoids the problem of what value to take for the particle diameter. Therefore it provides a useful basis for comparison of spherical and irregular particles.

In the past it has been difficult to study the effect of a wide particle size distribution because of difficulties with packing. Now that slurry packing methods seem to be reliable and less likely to lead to size segregation than dry packing, and closely sized spheres are available in the range 2–20 μm, the way is open for some detailed experiments to produce hard data on the effect of size distribution on the plate height–velocity function. From data on spheres may come an understanding of the effect of irregular shape combined with a wide size distribution. Current comparisons[109] suggest that shape per se is unimportant but there are other variations like pore size distribution and shape which make a direct comparison impossible at present.

The separation impedance concept also allows comparison of packed columns with other structures such as an open-pore polyurethane column described by Hansen and Sievers[379]. Their columns are indeed highly permeable ($K^\circ = 5.8\,\mu m^2$) but a little less attractive in practice because the efficiency for their impedance is less (E = 14 000) than for good porous silica columns (E = 2 500).

Flowrates giving Maximum Efficiency

By rearranging the reduced velocity equation $u = \nu D_m / d_p$ and assuming a porosity of 0.75; a particle diameter of 5 μm; reduced velocity of 5, an estimate of the highest speed leading to negligible increase in plate height; and diffusion coefficient 3×10^{-9} m^2/s,

then F_c (μl/s) $\approx 2 \times d_c^2$ (mm)

Flowrates calculated from this equation could form sensible nominal 'standard' rates for these column sizes, regardless of length, in the absence of some good reason to operate faster (or even better reason to operate more slowly). Variations in porosity or ±20% in diameter are unlikely to be significant, but lower values for water and other solvents may be appropriate because of lower diffusivity.

Standard flowrates for $\nu = 5$ with 5 μm particles

column bore	flowrate	
mm	μl/s	ml/min
2	10	0.5
3	20	1.3
5	50	3
8	100	8
23	1000	60

(severely rounded — plate height is not very sensitive to changes in ν)

Non-sorbed peak time

Another re-arrangement of the equation allows calculation of the time for the 'solvent front' t_o from $u = L/t_o$

$$t_o = L\,d_p/\nu\,D_m$$

If $\nu = 5$, a practical expression for calculating t_o is

$$t_o = \frac{L(m)\ d_p\ (\mu m)}{5 \times 3 \times 10^{-9}}$$ and a rule-of-thumb expression is L(cm) x d_p (μm)

Thus for L = 100 mm, and d_p = 5 μm, $t_o \approx 50$ s; for L = 250 mm, d_p = 10 μm, $t_o \approx 250$ s. This shows that when columns operating at their highest efficiency are compared, short columns of small particles give faster analyses.

Pressure requirements

If this expression for t_o is substituted in the equation for pressure drop

$$P = \phi' \eta L^{2^{\circ}}/t_o d_p^2$$

then $P = \phi' \eta L \nu D_m / d_p^3$

LC in practice P A Bristow hetp, Macclesfield UK ISBN 0 9504833 0 3

[Note that the surprising relationship $P \propto 1/d_p^3$ arises if the length L and reduced velocity are assumed constant. For two columns to have the same number of plates, one would reduce L with d_p to make the reduced length the same, when the relationship becomes the more familiar $P \propto 1/d_p^2$.]

Substituting some typical values

$$\phi' = 500, \nu = 5$$
$$D_m = 1.2 \times 10^{-9}\ m^2/s$$
$$\eta = 1 \times 10^{-3}\ N\,s/m^2$$

gives

$$P(bar) \approx 60\ L(mm)/d_p^3\ (\mu m)$$

Pressures needed for $\nu = 5$

type	d_p µm	L m	P[3] bar	P psi	t_o s
microparticle	5	0.1	50	750	33
'ultimate' see sec. 8.11	5	1	500	7500	330
popular	10	0.25	15.6	300[2]	170
pellicular	30	1	2.2	33	2000[4]
classical	100	0.5	0.03 (≃0.3 m head)	0.5	3300 (≃1 hr)

Note 1 Typically one would need $10 \times t_o$ for a complete analysis ($k' \simeq 10$).
2 Although a very low pressure drop is calculated, in practice most columns are operated at higher speeds (ν) trading for resolution.
3 The impedance of frits may be significant compared to that of the column.
4 Very much higher values of ν, typically ~100 are used.

The above expression can be used to assess the likely pressure requirements for various sized packings and column lengths. It is obvious that there is a limit to the smallest particles that can be used before the length becomes absurdly short; the ultimate limits are discussed further in section 8.11.

[Note that when a fixed number of plates per second are required a very different picture emerges — smaller particles then need *less* pressure. The condition here is that a reduced velocity of 5 is achieved.]

Or it can be reorganised in terms of volumetric flowrate. In this case a value for the porosity ϵ is required to calculate t_o from the equation

$$t_o = 0.785\, d_c^2\, L \epsilon / F_c$$

and hence the pressure drop P is given by

$$P = \phi' \eta L F_c / (0.785\, d_c^2\, d_p^2\, \epsilon)$$

For porous particles $\epsilon = 0.7$ and $\phi' = 500$ are typical, so a practical expression for P is

$$P(\text{bar}) = \frac{9.0 \times 10^6\, \eta\, (\text{mN s/m}^2)\, L(\text{m})\, F_c\, (\text{ml/s})}{d_c^2\, (\text{mm})\;\; d_p^2\, (\mu\text{m})}$$

Some typical results for a few types of column are tabulated below.

Pressure drop for various columns

eluent	η mNs/m^2	bore d_c mm	length L m	d_p μm	flowrate F_c ml/s	ml/min	pressure drop bar
hexane	0.3	5	0.1	5	0.02	1.2	8.6
methanol–water	1.8	3	0.25	6	0.02	1.2	250
chloroform	0.4	8	1	5	0.1	6	225
chloroform	0.4	23	0.5	10	1	60	34

This equation can be used to estimate whether the pipework of a system or the column end frits are causing excessive pressure drops.

Reduced Parameters and Irregular Particles of Varying Size

summary • **comparison of reduced parameters**

A practical difficulty in using reduced parameters arises when the particles are not spherical, or vary widely in size. What value for the particle diameter should be taken? Can spherical and irregular particle results be compared? This is certainly an extension to application of the theory and is open to criticism. Different answers result from different assumptions: these assumptions are not always stated explicitly.

Particle size could be assessed from the weight average but reduced length and height seem more naturally related to particle numbers. But the mean size based on a number average is often quite different from that derived from a weight average.

Alternatively the column impedance could be used to calculate effective particle size. On this basis irregular packings appear to have rather smaller mean diameters and thus larger reduced plate heights (by a factor of 2–5).

Wall effects and the 'Infinite Diameter Effect'

Knox, Laird and Raven[120] shed some light on our ignorance about wall effects, and their impact on the nett chromatographic result, by their fundamental studies, extending earlier work by Knox and coworkers[118,106]. They begin to map the distribution (and dispersion) of the solute as it emerges from the end of the column and show the existence of a marked wall effect where the plate height is three times as great. This region extends about 30 particles into the column (~0.65 mm with 21.5 μm spherical alumina). The interaction of this wall effect and the criteria for obtaining an infinite diameter column (and how it varies with linear velocity) are explored, and summarised by recommendation of the condition

$$(d_c - 60d_p)^2/L\,d_p \geqslant 16\ (1.4/\nu + 0.060)$$

for columns to be infinite diameter.

It is not clear how far these results can be extrapolated to more typical chromatographic systems with smaller particles, perhaps irregular with wider particle size distribution, and slurry packed: many observations of peak asymmetry—marked fronting—in small analytical columns suggest that these results are highly pertinent, both in trying to use the infinite diameter mode, and in the penalties of using columns in which these wall effects predominate and determine the observed plate height.

4 PACKING COLUMNS

The process of packing the separating material into the column is the subject of (too) much *mystique.* The objective is to reduce zone spreading, discussed in theory in section 3 and in practice here.

4.1 Dry Packing

summary

- **useful for larger particles (>30 μm) and smaller sphericals (>10 μm)**
- **still an art**
- **packing density useful criteria for large columns**
- **tap-fill method well proven for pelliculars**

Despite the vogue for slurry packing (and real advantages for small particles) dry packing is still best for some materials and column geometries. Large and/or spherical particles dry pack well. All apparatus and techniques involve vibrating the bed to settle the particles into as even and regular packing structure as possible, but without causing size segregation when the larger particles accumulate near the walls. Difficulty in describing and reproducing the direction and amplitude of vibration seems to be the explanation for the number of variations on this theme. Inter-laboratory comparisons, already difficult because of differences in hardware and packing material, and column assessment method, become impossible. Each lab must try as diligently as possible to follow a recipe but ultimately it comes down to the right touch. The time and expense of failures is depressing and accounts for the popularity of pre-packed columns; it will also add to the popularity of slurry packing if it is easier to follow a reliable recipe.

A basic problem is that there is no recognised way of assessing if a column is well packed until liquid is passed through. It is then annoying to have to recover and dry the packing in order to try again. There are some indications from work in the authors laboratory that packing density in g/ml may prove a reliable guide to good packing. The actual value of the target density will need to be determined for each type and size of material. Information gathered so far has come from packing of quite large diameter columns where accurate measurement of column dimensions and weight is much easier than for long narrow columns used to pack pelliculars.

Most packing gadgets lift the column with some eccentric cam arrangement and allow it to fall a set distance, thus attempting to standardize the vibration intensity. Vibrators of the type used in GC oscillating at line frequency are also useful though up and down vibration of a vertical column is better than horizontal vibration at the current packing level. Spherical particles with a very narrow particle size distribution (such as Spherisorb silicas S20W and aluminas A20Y) dry pack without size segregation problems using vertical vibration. Tolerable columns, especially with dense alumina, can be packed with 10 μm spherical materials. Rotation and gentle tapping or vibration at the current packing level have been used, in combination with vertical tapping down, to try to reduce the size segregation problem, and this forms part of most of the variants of the tap-fill method for pelliculars. Most manufacturers give details.

It is helpful to pack glass columns, for use or practice, to assess the amplitude of vibration and its effectiveness in settling the bed to the maximum packing density giving an even top surface, and to predict and verify the optimum feed rate of new material from above. Choice of wire size restricting a funnel orifice can control the feed rate. Coloured particles (sieved out samples larger and smaller from the main batch, dyed and dried) have been used to study size segregation. Armed with the right touch one can then pack metal columns, checking that the same packing density is achieved.

Dry Packing Method

(Based on the 'Tap-fill' method of Kirkland, reference 3, p. 189)

1 Clean the column if necessary with a chlorocarbon solvent, then scrub with a pipe cleaner and hot detergent solution, rinse with water and then acetone, and dry.
2 Attach the column end fitting.
3 Fix a small funnel to the top of the column with rubber tube.
4 Add enough material to fill the column a few mm (a few tenths of a gram for 2mm bore) holding the column vertical.
5 Tap the column firmly on the bench-top or floor a few times per second while rapping the side of the tube at the appropriate level with, say, a tube from an empty chart roll, rotating the tube slowly.
6 Tap several times gently without side rapping.
7 Add another portion and repeat until full.
8 Finally tap gently for a few minutes (without rapping on the side).
9 Remove a small amount of the packing at the column top and insert a metal or ptfe frit.

The process takes 20-40 minutes depending on patience and skill.

4.2 Slurry Packing

summary • **slurry packing best method but**
• **many different recipes and much conflicting evidence**

Packing from a liquid slurry has at last been accepted as *the* method for packing liquid chromatography columns. The liquid is needed to neutralize forces between particles. As the particle size falls the *external* surface area increases: typically for 5 μm spheres it is about 1 m^2/g. The surface forces, electrostatic in nature, cause agglomeration and prevent each particle falling freely into a regular packing structure that will give low zone broadening and high chromatographic efficiency. So slurry packing is essential for small particles. Slurry packed columns have also proved to be more permeable than those dry packed with the same material[109].

Organic resins for ion-exchange and permeation chromatography have always been slurry packed because they swell when wetted and dry-packed columns would block. As users of ion-exchange resins turned to smaller and smaller (and less closely sized) particles, the time for settlement of the slurry under gravity became too long and Scott and Lee[234] turned to dynamic packing, that is pumping the slurry into the column under pressure. This method was shown feasible for adsorbants like silica and alumina[232], but only taken up again by Majors[235] and many others when it was clearly impossible to dry-pack smaller particles.

(It is still traditional to dry-pack spherical, usually pellicular, material with size from 20 to 50 μm. Following the successful packing of 60–70 μm particles by Sie and van den Hoed[232] in 1969 with reduced plate heights of 2–4, slurry packing can just as well be used for pellicular 40 μm materials[236].)

A large number of variations in slurry packing methods have now been described using almost as many different versions of the apparatus. Areas of variation are:

- liquid composition (and whether a second immiscible liquid is used to expel the slug of slurry)
- slurry concentration
- reservoir shape and size
- pressure

The basic apparatus is:

Slurry liquid

Slurry liquids are chosen for

- low viscosity
- dispersing the particles to give a deflocculated suspension, or
- giving a flocculated suspension
- balancing the particle density so that particles do not fall under gravity

Suspension of particles has been studied in detail in connection with the formulation of stable suspension of pharmaceuticals[264] and glazes used in production of ceramics.

A flocculated suspension consists of aggregates of a few tens of particles. Visible characteristics are rapid settling with a clear boundary to give a settled bed of large volume that can easily be re-suspended. The ratio of the settled bed volume to the total slurry volume is called the sedimentation ratio. If equal to 1, the suspension is stable and will not settle further: this is usually the aim for a pharmaceutical product.

Deflocculated suspensions, by contrast, settle slowly approximately at the rate predicted by Stoke's Law with an indistinct boundary and leaving a turbid supernatant liquid, to a compact bed of low sedimentation ratio which is caked and difficult to re-suspend.

Thus one can assess the type of slurry from inspection of a small sample mixed in a test tube. For silica, water[178], methanol[174] and acetone[247] give de-flocculated slurries. Slurries with dense non-polar liquids like halocarbons, or hexane and ether, are well flocculated and their settling rates well above (≈50 times) that predicted by Stoke's Law.

But if a non-polar group like octadecylsilane is bonded on to make a reverse-phase material, polar liquids like water and methanol cause flocculation and settling rates are least in moderate polarity liquids like acetone, chloroform, carbon tetrachloride[265], or ether.

A low viscosity slurry liquid will reduce the time and/or pressure needed to pack a column. Methanol, acetone, ether, iodomethane[237], and hexane have particularly low viscosities, and cover a range of polarities.

Many workers have packed successful columns with both flocculated and deflocculated slurries. (Flocculation is usually a consequence of choosing a balanced density system.) But many authors also report less good results using other systems and it is difficult to draw any reliable conclusions from the evidence available at the time of writing and further research would be worthwhile. The author's preference is for deflocculated systems: for silica, methanol has proved reliable.

Acetone, a dipolar aprotic solvent, deflocculates both silicas *and* bonded-phase silicas[247].

Choice of slurry liquid may be influenced by the need to remove it before operating the column: haloalkanes are difficult to remove completely and interfere with use of short wavelength (<240 nm) detection. Methanol or water may be troublesome if dry alkane eluents are to be used.

Some workers use an immiscible liquid of different density to pump the plug of slurry from the reservoir into the column. This means that one can be sure that all the solid has been pushed into the column as soon as the liquid interface comes through. However when one knows the time and volume required for a given apparatus using only one liquid, this advantage is small.

Slurry concentration

There are wide variations in the concentration of solid in the slurry, usually determined incidentally by the reservoir geometry. It is not clear whether this makes any difference to the packing density or quality. Work by Bristow, Brittain, Riley and Williamson[174] and Webber and McKerrell[265] suggests that dilute slurries produce a more stable packing structure with spherical particles, but the evidence is not conclusive.

Reservoir

The simplest slurry reservoir is that described by Scott and Lee[234].

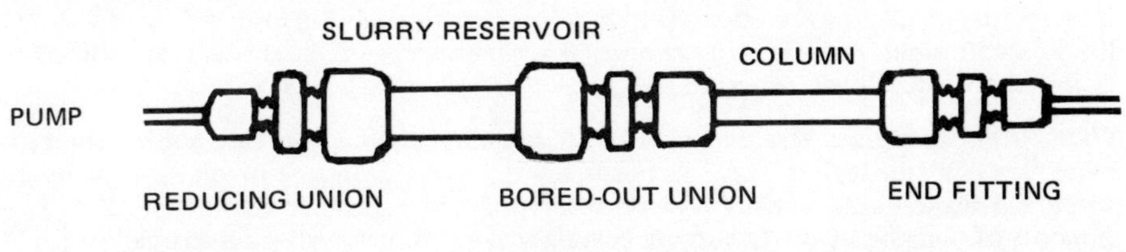

LC in practice P A Bristow hetp, Macclesfield UK ISBN 0 9504833 0 3

The tubes used for the reservoir may, or may not, be the same diameter as the column, and ratio of column to reservoir diameters from 1:1 to 1:10 have been reported. This apparatus has also been used to pack 1 in diameter preparative column with adsorbants of various sizes from 5 to 50 μm. Valves have been added to make it possible to add the slurry without disconnecting fittings but this is not essential.

If a larger reservoir is used a more dilute slurry is feasible. Sample cylinders are convenient and give a smooth transition from large to small diameter. Bored through fittings should be used to connect the column.

Godbille and Devaux[158,241] have used the column as its own reservoir and compress the slurry by using a jack to push the column end fitting up into the column bore. The liquid escapes through sinters at both ends. Despite the very low pressures used ($\approx$10 bar) and the wide variety of liquids used (the eluent to be used for that particular analysis) the columns are extremely efficient.

Linder, Keller and Frei[236] revive the system used by Sie and van den Hoed[232] but use much higher pressures. They feed slurry to the column from a stirred vessel into which fresh solvent is pumped. This produces a dilute slurry in a low viscosity liquid—they favour a dioxane-methanol mixture. Larger particles (30-50 μm) were also packed by this method.

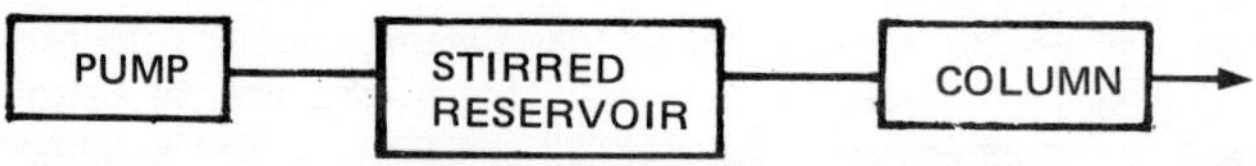

A similar effect can be achieved by upward packing[174] using a reservoir whose diameter is a few times greater than that of the column bore. This configuration dilutes the slurry as particles fall down the reservoir but are swept back up by the liquid flow. As particles are swept into the column their linear velocity suddenly increases and the high impact velocity helps good packing. An important practical advantage is that any lumps sink rapidly to the bottom of the reservoir and are not packed. A column 1 m long was packed with 3 μm silica particles at only 150 bar with this method[174].

Pressure

There are widely conflicting views on the pressure required for slurry packing—from $<$10 to near 1000 bar. In the light of this, the literature evidence on the effect of pressure is unconvincing. Since columns of 'theoretical' efficiency, that is reduced plate heights of $\approx$2 have been obtained by packing at almost any pressure in this range, the effect of pressure *per se* is probably small. If, as has been suggested[174], it is the particle's impact velocity that is important, then the viscosity of the slurry liquid determines the pressure needed. In general one must suspect that pressure requirements have been overstated.

When a liquid causes the particles to flocculate (usually the case with balanced-density silica slurries) it may be necessary to exert a higher pressure in order to force the agglomerates apart and re-order them in a closer packed structure. Proponents of balanced-density slurries certainly seem to use higher pressures.

The pressure requirement will obviously tend to rise as longer columns are to be packed. Many workers have reported that attempts to pack columns longer than 250 mm have generally been less successful: perhaps it is significant that two reports of success in packing longer columns[385,174] have specified deflocculated slurries. Wider columns also *seem* to be easier to pack, though again a column with a bore of only 1 mm and 1 m long has been 'well-packed'[385].

Despite the difficulties reported, it seems likely therefore that ways of packing longer columns with micro-particles will be found.

Slurry Packing Method

Apparatus

1 Pump, preferably constant pressure >100 bar with 1/16 diameter tube, 0.031 in bore to carry flow to reservoir. An on/off tap in the line may be convenient.
2 Sample cylinder (75 ml or 150 ml for large columns), Crawford Fitting Co 304-HDF4-75, Hoke International 304-4HD-75; with solvent feed line connected to the *bottom*, and a bored out fitting at the top to connect the column, Swagelok SS-400-1-4BT, Parker-Hannifin 4-4 FH4BZ-SS for 1/4 in columns; SS-600-1-4BT or 6-4 FH4BZ-SS for 3/8 in.
3 Column (1/4 or 3/8 in od) with end assembly.
4 1/16 od 0.032 in bore tube, ss or ptfe to take solvent to waste.
5 Funnel to fill sample cylinder—check it pours properly before trying slurry.

Method

1 Assemble column end fitting and swage nuts on top of column using the injector or other fitting to which it will be attached.
2 Charge pump with methanol or other suitable liquid.
3 Weight out silica (1.3–1.4 g per 100 mm of 5 mm bore tube).
4 Slurry silica in 70 ml methanol and shake well, perhaps ultrasonically.
5 Make sure everything is ready and spanners adjusted to correct size.
6 Shake slurry well and immediately pour into sample cylinder using funnel.
7 Quickly attach column *pointing up* and start flow judiciously in case anything is loose or leaking.
8 Pass at least 250 ml of methanol.
9 Turn off flow and wait a minute. Disconnect the column.
10 Remove top 5 mm of packing, push a 5 mm mesh disc into place, add glass beads and press ptfe sinter into top to hold beads in, and trim off a mm or so that remains proud.

The column is now ready for use as soon as any unwanted residual methanol is washed off by your eluent.

Packing large columns
(can be used for 3/8, 1/2 and 1 inch columns)

Apparatus

As shown on diagram. If the compression fittings for the column end are not standard depth types it may be necessary to use a 'bored-out' union. The tube to the pump should not be less than 0.8 mm bore (0.031 in) or rather much pressure drop will occur. A valve in the line to the pump is desirable to avoid any chance of packing material getting back into the pump valves. The column can be used to hold slurry too if the column end is removed to pour slurry in. Whether or not this is convenient will depend on type of fitting.

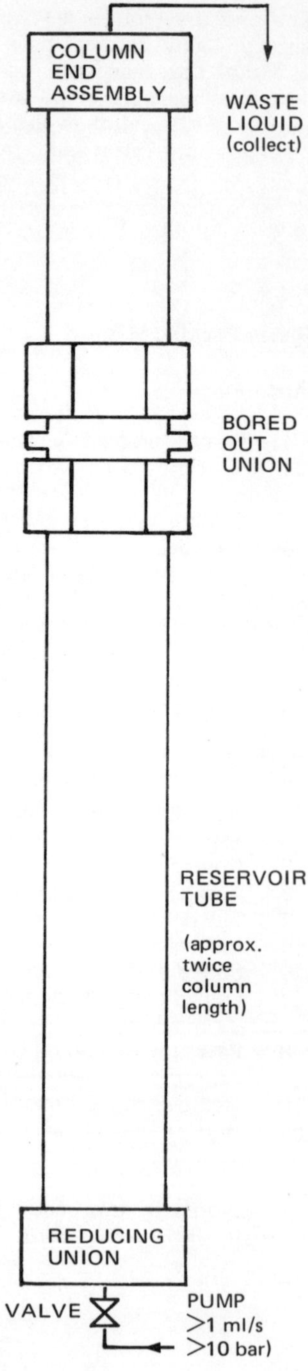

Method

1 Load the pump with methanol or other suitable liquid for slurrying the particular packing. Turn the valve off.
2 Calculate or measure the volume of the reservoir tube (and the column too if this is also to be used to hold slurry). For example a 1 inch tube 1 metre long will have a volume of ~400 ml.
3 Weigh out the adsorbant and make a slurry with methanol, or other suitable liquid, whose final volume is a little less than the reservoir volume. A measuring cylinder with a plastic stopper is most convenient. Shake well and check for even dispersion and freedom from lumps.
4 Clamp the reservoir tube pointing vertically upwards.
5 Check that everything is ready for the next steps and that spanners are available and set to size if necessary.
6 Re-shake the slurry and pour in.
7 Attach the column or column end assembly and tighten. Connect the waste pipe.
8 Turn the pump on and collect at least twice the reservoir volume, depending on length, particle size and pressure available. Use as high a pressure as is safe for the tubing and fittings.
9 Turn the pump off and wait until no more liquid emerges from the waste pipe. This is important and usually takes a minute or so.
10 Disconnect the 1 inch union and remove the column from the reservoir, and make any special column top arrangements.
11 Connect to the injection system and the column is ready to use after any equilibration with the eluent to be used.

LC in practice P A Bristow hetp, Macclesfield UK ISBN 0 9504833 0 3

5 THEORY OF RETENTION AND RESOLUTION

So far the particles have been assumed inert. Useful materials adsorb, desorb and thus separate. Ideally this process should not slow down the equilibration process more than eddy and molecular diffusion, and therefore not add to the observed zone spreading. In practice, of course, it does. For some materials like ion-exchange resins it adds very greatly. In general thin layers, only a few molecules thick, such as a layer of water or adsorbed component of eluent, add little, but thicker layers commonly used in pellicular packings or thick liquid films, can increase dispersion and thus peak width, mainly because molecules take much longer to diffuse in and out of this thick and often viscous (low diffusivity) layer.

Packings today give higher efficiency and narrower peaks because of improvements in the rate of equilibration, that is fast mass transfer.

Further improvements can be expected.

Partition and Adsorption

Partition is a term used to imply a distribution between two phases depending on solubility. In GLC, and in true LLC, the term has some meaning but the distinction between adsorption and partition is not very useful for practical materials in use for LC today. The observed parameter is the amount a compound is retained. Hanai and Fujimura[301] suggest that the term 'adsorbed' be used without any judgment about the mechanism for both organic and inorganic adsorbents and this convention will be followed here, with the term retention implying the result of adsorption.

An example of a case where partition coefficients themselves are the object of interest is in developing biological correlations. a technique pioneered by Hansch[294], and now an area of rapidly growing interest. One main barrier to widespread use of these correlations to predict the action of drugs has been the difficulty of measuring partition coefficients except by tedious and error-prone methods: this has now been removed by the rapid measurement of retention on a suitable chromatographic column, even on quite impure samples. Useful results are already emerging[295–298] but it has been argued[295] that it is much safer to use a true liquid—liquid octanol—water system rather than octadecylsilane bonded-phases, a convenient but questionable alternative.

Retention

summary

- **k′ is number of column volumes needed to elute a solute**
- **k′ is linear with time, volume and chart distance**
- **retention ratio R is linear with distance on column or TLC plate**

In order to separate (except by size alone) solutes must be retained by the column—eluent system.

Two equivalent methods have been devised to express the amount of retention and yet be independent of column size, flowrate, time, and chart distance.

k' — column capacity ratio and retention ratio

The so-called 'column capacity ratio', now almost invariably called simply k' (because the term is so long-winded, un-informative and ripe for superannuation), is defined in theory as the ratio of the amount of solute in the stationary phase compared to the mobile phase.

Its practical usefulness is as a *retention parameter*, corrected for the volume of the spaces between and pores inside the particles, and linear in time, volume and chart distance (sometimes spoken of colloquially as the 'column volume' V_M). A permeating but non-sorbed solute has $k' = 0$; k' increases by 1 for each 'column volume' needed to elute the compound.

$$k' = (t_r - t_o)/t_o = (V_R - V_M)/V_M$$

It may be useful to note that the retention volume $V_R = V_M (1 + k')$.

Algebraically related to k' is the *retention ratio* R, the fraction of solute in the mobile phase

$$R = 1/(1 + k') = t_o/t_r = V_M/V_R = z_o/z_r$$

where z_o is the distance travelled by the solvent, and z_r the distance travelled by the solute zone *on a column, or plate.*

This retention parameter is linear in distance eluted on the column, or TLC plate where the closely related parameter R_f is generally used.

Both these parameters can be determined by direct measurement from either a chromatogram or a column, or plate. k' is normally used to treat results from columns: R (or R_f) for TLC and paper chromatography.

Retention ratio R versus Mass distribution ratio k'

k'	R	R_f
∞	0	0
20	0.05	0.04
10	0.09	0.08
5	0.17	0.14
4	0.20	0.17
3	0.25	0.21
2	0.33	0.28
1	0.50	0.43
0	1	0.85
$k' = (1 - R)/R$	$R = 1/(1 + k')$	R_f is ~15% less than R measured with TLC and Paper chromatography[9]

LC in practice P A Bristow hetp, Macclesfield UK ISBN 0 9504833 0 3

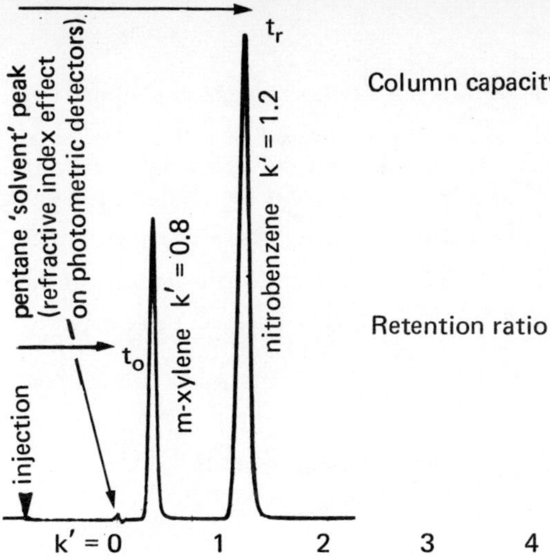

Column capacity ratio $k' = \frac{\text{amount in stationary phase}}{\text{amount in mobile phase}}$

$= (t_r - t_o)/t_o$

Retention ratio R = fraction of solute in mobile phase

$= t_o/t_r = 1/(1 + k')$

one column volume is
about 3/4 of internal volume of tube

Separation

summary
- **zone width increases with $\sqrt{L}$**
- **separation of peak centers $\propto L$**
- **but selectivity most important for practical separations**

Since zones spread steadily as they pass down the column, it is not obvious why separation of solute zones is ever possible. Fortunately, while the zone width (4σ) increases with $\sqrt{L}$ (and with $\sqrt{n}$ and $\sqrt{H}$) the difference between zone centers is directly proportional to L.

In theory then a sufficiently long column will always provide enough separation. In practice the pressure required may be too great, or the peaks will be too dilute to detect when they do elute (and there are some problems in packing very long columns efficiently). Hence the need for efficient columns which only spread the solute zones slowly continues — and for highly selective separating materials because it is always easier to separate compounds which are *selectively* retained by the column packing material. Classical column separations have always relied on good selectivity, because so few (a hundred or two) plates could be produced with large particles and dry packing methods.

Resolution

summary
- **R_s quantitative measure of separation**
- **peak shape important in practice**

Resolution is a measure of how much two components are separated. It is defined by

$$R_s = \frac{\text{difference in retention times}}{\text{mean peak (base) width}} = \frac{t_r^A - t_r^B}{\frac{1}{2}(w^A + w^B)}$$

LC in practice P A Bristow hetp, Macclesfield UK ISBN 0 9504833 0 3

For well-retained components, the peak widths are similar enough to simplify this to

$$R_s = (t_r^A - t_r^B)/\text{basewidth } (4\sigma)$$

If $R_s = 1$, the peak maxima will be one base width (4σ) apart, and the overlap ~2%, and so reasonably well resolved for most purposes.

$R_s = 2$ represents an excellent resolution with 2 base widths between peak maxima, and one base width of near-baseline. In theory the overlap is less than 0.1%, but in practice distortions of the peak shape (tailing and fronting) determine the effective resolution.

But if R_s is only 0.8, or less, the peaks are normally considered to be only partially resolved, and accurate division of the total peak area between the two peaks is difficult.

summary	• R_s <0.8	**partial resolution — some overlap**
	~1	**normal criterion — small overlap**
	~2	**excellent separation — no overlap unless peaks tail**

Assuming symmetrical Gaussian peaks with equal base widths, resolution can be shown to be a function of three factors, of which two are often combined

$$R_s = \frac{\tfrac{1}{4} \cdot \Delta k'}{k'} \cdot \overbrace{\frac{k'}{1+k'} \cdot \sqrt{N}}^{\text{effective plates}}$$

relative difference in retention — selectivity; degree of retention; column efficiency

It should be immediately clear that the most important factor determining resolution is the selectivity of separation ($\Delta k'$) by the column–eluent system, expressed by the term $\Delta k'/k'$, which is equivalent to

$$\frac{\Delta t_r}{(t_r - t_o)}$$

or in terms of relative retention a where

$$a = k'_A/k'_B = (t_r^A - t_o)/(t_r^B - t_o)$$

this term is equivalent to $(a - 1)/a$.

The column efficiency factor $\sqrt{N}$ is quite important but one must not forget the square root relationship which means that only large improvements (say 30%) are significant in practice. Nevertheless improvement from the 100 plates typically obtained from classical columns to 100 000 feasible with very high efficiency columns has given a very significant 30-fold improvement in resolving power (and at least a hundredfold improvement in speed).

The degree of retention factor $k'/(1 + k')$ needs considering mainly when k' is small enough that k' differs appreciably from $1 + k'$, say $k' \simeq 3$. As peaks are more retained, the improvement in resolution is much slower, and eventually becomes negligible (and detection more difficult). If components are retained less so that $k' < 1$, the resolution falls sharply.

summary
- **selectivity or relative difference in retention is most important factor**
- **solutes must be retained to be resolved [$k' > 1$]**
- **rarely worth increasing retention so that $k' > 20$ to resolve two close peaks**
- **plate number important (but depends only on $\sqrt{N}$)**

Effective plates

summary
- **$k' \approx 2$ gives highest speed of resolution**

The concept of effective plates N_{eff} incorporates the second two factors in the resolution equation above and has been used as a performance index which takes account of the degree of retention. An additional step is to estimate the speed performance by calculating the effective plates per second N_{eff}/t_r.

One expression for this parameter is obtained by noting that $t_r = t_o\ (1 + k')$, thus

$$\frac{N_{eff}}{t_r} = \frac{N}{t_o} \cdot \frac{k'^2}{(1 + k')^3}$$

Apart from showing that separation is faster the higher the flow velocity u and the shorter the unretained peak time t_o, it is interesting in suggesting the most useful range of k' to select by adjusting the eluent strength in order to maximise the speed of resolution. This function shows a steep rise from when k' is near zero, to a maximum at $k' = 2$, and then a fall at higher k'.

Thus the most practical range for k' is 1—10 and for highest speed 1.5—3. Of course this assumes an ideal separation of a few components retained by conveniently similar amounts: practical problems often require the use of a larger range of k' to resolve two rather similar components from others that are very different, or simply to resolve more components, that is to increase the peak capacity.

Peak Capacity

summary
- **number of equi-spaced peaks a column could theoretically separate**
- **roughly $\frac{1}{2}\sqrt{N}$ for $k' \simeq 10$**
- **gradient elution necessary for more than a few hundred peaks**

The number of peaks which can be fitted onto a chromatogram (with all bases (4σ) just touching and $R_s = 1$) was introduced by Giddings as the peak capacity and defined, with some simplifying assumptions, as

$$\phi = 1 + 0.6\sqrt{N} \cdot \log\,(1 + k')$$

In practice we may approximate this expression, for an arbitrary choice of $k' = 10$ for the most retained component, to

$$\text{peak capacity} \approx \tfrac{1}{2}\sqrt{N}$$

Thus for traditional large particles with $N \simeq 100$ only a few components can be separated. With typical HPLC columns giving $N = 10\,000$, about 50 compounds might be separated. By extending the maximum retention, the number of components can be increased, but only by $\log(1 + k')$, which also increases the analysis times and solvent consumption. In practice gradient elution is the only sensible way to increase peak capacity beyond about 300 ($N = 10^5$, $k' = 30$).

Efficiency in chromatography

Delley[100] has recently proposed that plate number and height concepts be abandoned, and that 'efficiency' be defined in terms of the relation of migration distance to zone (base) width. That is

$$\text{efficiency } E = L/4\,\sigma_L = V/4\,\sigma_V = t/4\sigma_t$$

Basing equations on standard deviations (σ) rather than variances (σ^2) does lead to some simplifications, thus resolution is given by

$$R_s = E\,\Delta t/t_r$$

and efficiency as a function of column length is given by

$$E_2 = E_1\sqrt{L_2/L_1}$$

and emphasis that separation efficiency is only the square root of 'plate height efficiency', a point sometimes forgotten.

LC in practice P A Bristow **hetp**, Macclesfield UK ISBN 0 9504833 0 3

6 SEPARATION MECHANISMS AND MATERIALS

summary

- **separation mechanisms are complex in LC**
- **liquid eluent is much more important than mere carrier gas**
- **more than one mechanism common**
- **eluent often modifies solute by ion-pairing or solvation**

For all our use of chromatographic materials to separate different molecules, and many studies and rationalisations of observed results, there is still quite an element of mystery about what really happens in the pores and on the surface of the column packing during separation.

The conventional wisdom is that the separation mechanism can be neatly pigeon-holed into liquid-solid (adsorption) chromatography (LSC), liquid-liquid (partition) chromatography (LLC), ion-exchange chromatography (IEC), and (gel) permeation chromatography (PC). This framework will not be abandoned but it does represent a considerable oversimplification that the user must be aware of.

Compared to GC, the mechanisms of separation in LC are much more complicated. The reason is simple: in GC the mobile phase is an inert gas, typically helium, whose molecules are very much smaller than the molecules to be separated. In gas-liquid chromatography, partition takes place between the gas (mobile) phase and the liquid (stationary) phase by a competition between solubility of the sample in the liquid phase and volatility which evaporates it into the mobile phase. This process is more or less independent of the carrier gas. An established theory attests to our understanding of the mechanism.

Eluents in LC are much larger molecules, more similar to the molecules to be separated, and they interact strongly with the separating surface, whether it be liquid or solid, or in between. Changing the eluent can change retention by several orders of magnitude, often much more than any change in the separating material can cause. (In contrast with GC, changing the temperature has only small, but subtle, effects.) Inevitably the process of separation is very complex. This complexity also means that there is much more scope for highly specific separations that are quite impossible with GC.

[In recognition of the role of the liquid, the term eluent is preferred to mobile phase, and often the term surface will be used instead of stationary phase to describe the zone of solid and/or liquid (which may be quite thick) where separation takes place. Larionov and Myers[384] point out that the *"microscopic dimensions"* of the *"adsorbed phase"* are too small for it to have real thermodynamic properties. For this reason the term 'zone' often seems more appropriate than 'phase'.]

To put the problem in perspective it may help to look at the molecular scale. Pores in typical silicas are around 10 nm. Thus the distance across the diameter is only a few score of solvent molecules. The zone over which the structure of the liquid is perturbed by the surface extends very roughly ten molecules deep and this zone occupies a substantial fraction of the total volume of the pore. So the structure and composition of liquid in any pores, or near the surface of a non-porous bead may be quite different from that of the bulk eluent in the much bigger interstices between particles.

In practice the following scheme of separation mechanism may be more helpful. (An indication of how the traditional classifications fit in is given.) Retention results when solute molecules are more strongly attracted to the separating surface zone than the mobile eluent.

Separation mechanisms

mechanism	**example**	**notes**
Size (exclusion and permeation)	GPC	How easily do molecules permeate the pores? Only if they permeate can they be retained by other mechanisms.
Retention by **surface**	LSC on inorganics like silica and alumina, or carbon (?)	Surface hydroxyl groups most important. Rare in practice, e.g. aromatic hydrocarbons with dry alkane eluent.
Retention on **surface layer** of *reversibly* bonded molecules	polar eluents used with silica, perhaps also with a bonded organic layer, soluble in eluent	Most practical chromatography. Typical example is use of water or methanol in hexane on silica.
Retention on *thin* **surface layer** of *permanently* bonded molecules	chemically bonded like ODS on silica, (often inadvertently)	Surface is often non-polar, but may be very specific like affinity chromatography, or groups can be polar ion-exchangers.
Retention on *thick* **surface layers**	polymers bonded to solid beads. Thick layers of viscous liquid insoluble in eluent (LLC). Solid resin beads (IEC and GPC)	Thicker layers give slower mass transfer so these are less attractive materials, though they may separate well, and have high capacity. If thick enough we may count the whole matrix as a very thick layer.

The distinctions between these classes are tenuous. There is the problem of the polarity of solvent at which the eluent is sufficiently polar to interact with the surface. When does one classify a layer as thick? Many bonded phases leave some exposed surface hydroxyl groups. How does a material like carbon fit into this classification?

It is also important to note that the eluent nearly always modifies the solute molecules in some way, at least by solvation, but more significantly by forming an ion-pair, most commonly by protonation or deprotonation. This is dicussed in more detail in section 8.10 on derivatization but needs constantly to be kept in mind when planning or rationalising a separation.

Reverse phase – non-polar adsorbents

Reverse phase is a term which could well be retired gracefully: only the 'column capacity ratio' is more confusing to those new to chromatography. The term arose from the use of hydrocarbon or silicone coating to make materials water repellent.

In a reversed-phase system the eluent is polar and the stationary phase is non-polar. (Thus it is the reverse of the normal phase system where the stationary phase is polar.) Thus non-polar molecules are attracted and retained. Making the eluent more polar has the reverse effect compared to normal systems and non-polar compounds are more retained.

The most non-polar phase devised is carbon[137] which promises to be a most useful material for solving many problems in LC. Silica is the most popular polar adsorbent.

silica	*polar* phase	polar molecules adsorbed most strongly	polar liquids elute most strongly
carbon	*non-polar* phase	non-polar molecules most strongly adsorbed	less polar liquids elute most strongly

Practical reverse-phase materials are nearly all bonded phases like octadecylsilane (or thick layers of coating liquid), but many resins used for permeation chromatography and ion-exchange chromatography also have considerable non-polar adsorption power if used with the right eluent. Many separations do not depend entirely on their sizing or ion-exchanging properties.

An interesting example of adsorption of sulphur S_8, a highly non-polar molecule, being retained on a non-polar styrene-divinylbenzene co-polymer, is described by Cassidy[299]. The sulphur permeates this large pore gel designed for GPC, while large organic molecules are excluded and small organic molecules are too polar to be retained. If the eluent is made more polar, the retention of sulphur is increased.

eluent		k' for S_8
tetrahydrofuran		0.28
tetrahydrofuran–methanol	(90:10)	0.49
	(75:25)	0.93

Another example of retention of non-polar compounds is provided by Nakae and Muto[300] again using a styrene–divinylbenzene co-polymer to separate alkylbenzene (and benzoates). Their results parallel those of Sleight[302] using ODS-Permaphase, a thick layer of polymerised silicone on a solid bead, who showed that the retention of alkylbenzenes is increased with increasing carbon number, unlike normal phase where all elute together. Sleight suggests that this is because the increasing alkyl group size makes the molecule less soluble in the polar (methanol–water) eluent.

For polynuclear aromatic hydrocarbons, however, the more rings the more retention. This retention order is the same as for polar adsorbents like silica but presumably it is for different reasons.

Sabbagh and Fagerson[376] concluded that a very different separation, glucose oligomers on polyacrylamide with pure water as the eluent, was not based only on sizing but also on the solubility of the carbohydrate in the organic resin.

A final complication is that in most practical applications more than one mechanism is responsible for the observed separation.

In this section, as befits a practical text, the classification is made on the basis of the type of material in use, with a discussion of the mechanisms of separation that are common with these materials, and some of their properties, advantages and problems.

section

6.1 Structure of adsorbents, silica, alumina, and carbon, and separations based on thin reversibly adsorbed layers, usually a component of the eluent.

6.2 Bonded-phases, whether thick or thin, or the whole matrix of organic material, including the very specific bonded materials used in affinity chromatography.

6.3 All materials which have ion-exchanging properties, and separation mechanisms on these, even if this is not specifically ion-exchange.

6.4 Size separations — permeation chromatography.

6.5 Pellicular materials.

6.1 Adsorbents

Silica

summary • **the most useful chromatographic material**
• **hydroxyl groups on surface cause adsorption**

The important properties of chromatographic silica (and alumina) have been described by Halpaap[323,333] together with a summary of the methods of measuring the various parameters. Although much of this data was collected for use with thin-layer chromatography it is equally valid for LC in columns under pressure.

Kohlschutter and Unger[9] also give a concise description of chromatographic silica gels.

Silica gel has the composition $SiO_2\ xH_2O$. (The term silica, silica gel, silicic acid are practically synonymous and only silica will be used here.) The material can be derived, at least conceptually, by polycondensation and dehydration of silicic acid. The equation below shows how two molecules eliminate one of water and form an Si—O bond.

$$2\,Si(OH)_4 \longrightarrow (HO)_3Si{-}O{-}Si(OH)_3 + H_2O$$

A network of Si—O bonds is built up by dehydration until tetrahedra of SiO_4 are linked together with most remaining hydroxyl groups on the surface.

As more and more water is lost the material changes from a liquid-like sol in which there are less-hydrated aggregates of silica 0.5-1 nm in diameter to a jelly-like gel when the forces between these aggregates have become large enough to give the structure a distinct shape. If this gel material is heated towards 200° C all physisorbed surface water is removed leaving a surface which is fully hydroxylated. At higher temperatures these groups are lost and at ~500°C only siloxane groups are exposed at the surface. At still higher temperatures the aggregate begins to sinter and both pore volume and particle size are reduced. Finally at ~1500°C the material fuses to give the transparent glass used for many optical elements in scientific instruments. (Reference 24, C Okkerse, chap. 5, p. 213).

Scott and Kucera[337] and Gray[284] describe the chromatographic behaviour of commercial silicas heated up by stages to 1000°C. The results of these experiments, and those using other techniques of assessing the hydroxyl groups per unit area[352], or weight, do not entirely agree with each other but this is hardly important in practice. Dehydroxylation can be reversed (provided the pores have not actually sintered) by reacting with water. This is a slow process under typical non-aqueous chromatographic conditions. To obtain a material that is most suitable for use as an adsorbent, heating a fully hydroxylated silica to between 200 and 400°C produces the most useful material.

The silanol groups are slightly acidic (about as much as phenol pKa ≃9)

$$Si{-}OH \rightleftharpoons Si{-}O^- + H^+$$

so basic compounds are slightly more adsorbed than neutral or acidic compounds (though many eluent systems drastically alter this simple mechanism). This means that silica itself can act as a cation ion-exchanger[336]. This may not always be immediately recognised as a possible separation mechanism. The activity of the silica and its determination are also discussed by Halpaap[323] and Snyder (reference 8, Appendix II) but has become less relevant to practical operation with many solvent systems which reach equilibrium with the surface and contain at least one highly polar component.

Surface of adsorbents

If the size and shape of the outside of the particles and the pore structure inside is complex, the surface and its interaction with the solute and eluent is more complex still. Much has been written about the surface, in particular by Snyder[8] from a chromatographic point of view, and by many others with other interests like catalysis. Kiselev and his school have made an extensive and fundamental study of adsorbents: their work is reviewed in reference 26.

Most of the chromatographic properties of the surface derive from the silanol groups on the surface, but siloxane groups may also make some contribution[335]. Snyder[8] has discussed the evidence for the different properties of different types of silanol groups.

Viewed from above the surface of a typical silica surface will appear thus, except that each silicon has a free vacancy

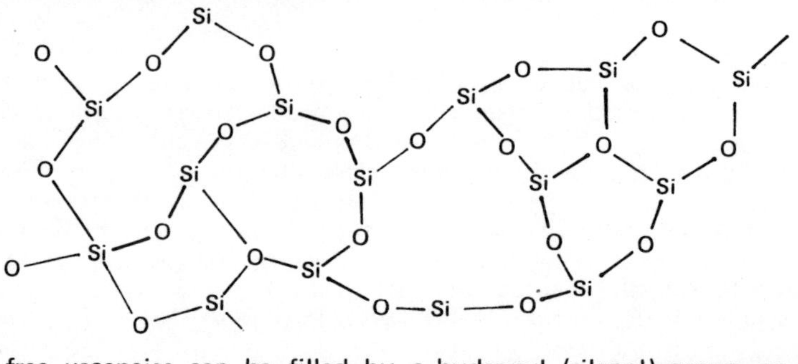

These free vacancies can be filled by a hydroxyl (silanol) group, or a siloxane bridge, (or be linked to atoms deeper into the surface). Hydroxyl groups may, or may not, be adjacent to another hydroxyl and thus be able to hygrogen bond together.

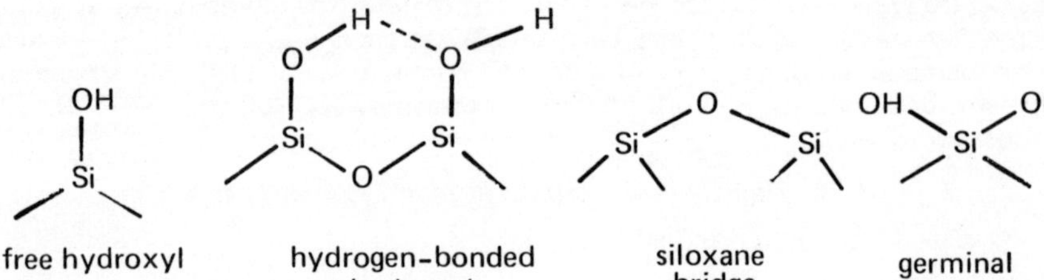

free hydroxyl — hydrogen-bonded hydroxyl — siloxane bridge — germinal hydroxyl

Water can hydrogen bond, for example thus

$$
\begin{array}{c}
| \\
-\mathrm{Si}-\mathrm{O}-\mathrm{H} \text{ ------ } \mathrm{OH_2} \\
|
\end{array}
$$

The calculated maximum density of hydroxyl groups is about 8 per nm^2 (1 nm^2 = 100 $Å^2$) but in practice most determinations give ~5 OH/nm^2 for fully hydroxylated silica, of which 1–2 are free and 3–4 are hydrogen-bonded to each other (assessed from IR spectra[335]). Full hydroxylation requires heating with water at about 100°C for some hours: this process is a chemical reaction, unlike physical adsorption of water.

If we consider a section of pore with diameter ~10 nm and 1 nm long, the surface area is ~30 nm^2 and there will be about 120 hydroxyl groups. If the specific surface area is 200 m^2/g there will be nearly 10^{21} hydroxyl groups per gram of column packing.

Controlled-pore glasses

summary
- **very small pore size distribution**
- **wide range of pore sizes**
- **granular**

These materials are made by an entirely different process from the silica gels. A glass is made in which there are crystalline zones of material which can be leached out leaving a material which is nearly all silica with closely sized and interconnected pores. An extensive bibliography[328] is available from Electro-nucleonics who now hold the licence from Corning, who first produced the material. Pore sizes between 7.5 and 200 nm are available in 8 grades. When ground and sized to 10 µm, Knox[109] found that it was an outstanding material for HPLC, as well as a material which has been found suitable, in larger mesh and pore sizes, for size preparations of biological molecules. Holdoway[99] reported that this material had the smallest pore size range of all the materials he studied (which included many in current use).

As with silica, adsorption is a complication of separations based on size. Silianization is recommended for use with non-aqueous systems, for example when separating synthetic polymers like polystyrene, and also polyethylene glycol 20M treatment, which forms a surface film, for aqueous separations of biological molecules[334]. Glyceryl-CPG, a bonded-layer material is reported to be superior to a non-bonded layer for many separations[366,371].

Alumina

summary • **alumina similar to silica**

Alumina is formed in a similar way to silica by condensation of hydrated aluminium cations (reference 24, B C Lippens, chap. 4, p. 171).

$$2\,Al(H_2O)_6{}^{3+} \longrightarrow (H_2O)_5\,Al{-}O{-}Al\,(OH_2)_5{}^{4+} + 2\,H_2O$$

thereby forming an Al—O—Al bridge.

Alumina consists of a mixture of hydrated aluminium oxides. The overall composition is about $Al_2O_3 \cdot \sim 0.5\,H_2O$.

Alumina has a similar, but subtly different, structure to silica. It exists in many crystal forms, labelled by greek letter prefixes, but which need not concern us further here: γ-alumina is the most common form used for chromatography. The surface structure has been extensively studied by Peri[287]. The surface hydroxyls are not completely removed even at 1000°C, but water only reacts slowly to form hydroxyl groups from adsorbed water at room temperature. Unlike silica, the aluminium ions are believed to play an important part in separation.

The complexity of the processes of dehydration and rehydration, and chemisorption of water onto the surface of alumina is well illustrated by the work of Borello and co-workers[343]. Spectroscopic and other methods give different, but complementary, information about water and hydroxyl groups.

Very few applications have been reported on porous aluminas though some types are now available in very small particle sizes (for example Spherisorb A5Y, or Merck Lichrosorb Alox T) even though some of the most efficient columns have been packed with these materials[120]. Alumina (Spherisorb) has also been used as a basis for a bonded-phase material[237]. That they give similar separations to silica but cost more is one reason why they have not been used as widely as silica. However some slight differences in selectivity are certainly expected on the basis of evidence from separations achieved on larger materials in traditional column chromatography. So alumina can be expected to give some useful separations which fail on silica. In their enthusiasm to achieve separations with silica, practising chromatographers may have unfairly ignored this material.

Methods of making porous materials for chromatography

summary • **angular material made from sodium silicate and coarse lumps of gel crushed and sized**

There is an element of mystery shrouding the method of particle production, perhaps more than is necessary for commercial and marketing reasons. From the academic and patent literature, and information released by manufacturers, enough information can be gleaned about the structure, properties and methods of manufacture to satisfy most users needs, and curiosity. All micro-porous materials for HPLC now on sale are chromatographically satisfactory, and the difference between them quite minor. There are no magic ingredients.

The method of making most silica used in traditional column chromatography and TLC is outlined below. This basic process, though different in detail, is used to produce Partisil and Lichrosorb, two popular irregularly-shaped (angular) silicas.

Dilute hydrochloric acid is added to a solution of sodium silicate. First a sol is developed and then a hydrogel forms. The pH at which gelation occurs strongly affects the surface area. Ageing and washing develop the pore structure, increasing pore volume and decreasing surface area, usually down to $\sim 400\ m^2/g$. Control of the production variables allows manufacturers to produce a material with suitable pore size and shape, but this procedure produces coarse lumps of silica which must be ground up and the fine particles sized.

Commercial materials are usually sized with an air classifier like the Alpine Multiplex[19]. (Hydraulic sizing or elutriation[19,21,251] can give closer sizing but it is rather expensive for a commercial product. It may be possible to improve air-sized materials a little by some sedimentation procedures as suggested by some workers, but losses are usually significant, and gains marginal.) The air classifier has a rotating wheel with zig-zag shaped channels: separation is achieved by balancing centrifugal and air-frictional forces. Cut sizes between 1 and 100 μm can be made. Naturally there is a trade-off between cut precision and production rate. The larger sizes are normally collected first. Smaller sizes are more expensive because more passes through the sizer are needed.

Typical particle size distributions for Partisil are[320]:

% oversize	Particle Size		
	Partisil-20	Partisil-10	Partisil-5
1	35.6	25.6	10.9
5	30.1	12.2	9.3
10	27.6	11.6	8.7
50	20.7	9.8	6.3
90	17.5	7.8	3.6
95	14.5	7.0	3.0
99	10.9	3.8	2.2

The surface of silicas can be modified after manufacture by, for example, boiling with dilute acid to fully hydroxylate the surface, or by heating to remove adsorbed water, or even the hydroxyl groups. The whole pore structure can also be altered by more drastic treatment, either chemical or physical. Aue and Kapila[326] discuss the modification of silica for chromatography. Heating with water at about 400°C has been the subject of extensive studies by Kiseley and co-workers. (For example reference 327, or references 12-27 given by Aue[326].) This 'hydrothermal' treatment drastically increases the pore size with a corresponding reduction in surface area but has little effect on the pore volume.

Spherical particles

summary
- **spherical particles made by gelling droplets of sol**
- **pore structure results mainly from sol particle size**

Unger[303] describes the preparation of spheres by hydrolytic polycondensation of polyethoxysilane. A viscous sol of partially condensed ethoxysilane is stirred vigorously to form an emulsion with a water–ethanol mixture, and an added catalyst causes the spherical droplets in the emulsion to gel. Many variables control the particle and pore size. Lichrospher materials (E Merck) are presumably made by this method[322].

Kirkland[342] has described some silica microspheres now sold (in packed columns) by Du Pont as Zorbax-SIL. A related patent[325] describes the manufacturing process a little. The aggregates of silica in the sol stage are around 5 nm. The pore structure is created by mixing in urea and formaldehyde. These polymerise and this viscous mixture is stirred vigorously to form droplets which gel separately. After washing the material is heated to burn out the organic polymer[389].

It is believed that Spherisorb chromatographic materials are made by a more refined sol-gel process. Such processes have been the subject of detailed research and patent action because of their application in the preparation of nuclear fuel particles. These particles should be free-flowing spheres which can be packed into metal cans to a controlled density, and because of their toxicity the powders must be free from dust. Spherisorb chromatographic materials are a spin-off product from the atomic energy programme at AERE, Harwell. The principles have been described by Fletcher and Hardy[289] and Dell[340] and patents reveal some of the practical complexity. The details of the particular experimental procedures used are not available at present but the route evidently affords a close control of the internal pore structure of the spheres. A comparative study[99] of the pore size distribution in Zorbax (Dupont), Lichrosorb (Merck), Partisil (Whatman) and Spherisorb chromatographic materials reveals that the latter has the narrowest distribution with no pores below 7 nm or above 13 nm in diameter (see Fig. 8 or ref. 99).

Porosity

summary
- **columns porous between particles**
- **and inside pores of particles**
- **observed total porosity includes both**

Chromatographic beds are porous because of the spaces between the particles: most materials are also porous because there are holes *inside* the particles. Thus there are three distinct definitions of column porosity – which leads to confusion, even in communications between experts.

Porosity is the fraction of space: for a column a value of 0 means that it is solid and non-porous, and 1 an empty tube.

If the packing material is solid (or considered as if it were solid) we can define the interstitial or interparticle porosity as

$$\epsilon_I = \text{volume of spaces between particles/volume of column tube}$$

The porosity of individual particles is easy to visualise (but difficult to measure directly and unequivocally)

$$\epsilon_{particle} = \text{volume of pores/total volume of particle}$$

When we consider chromatography of columns packed with porous particles when all molecules are small enough to permeate the pores, the observed parameter is the total porosity

$$\epsilon_{total} = \text{(volume of spaces + volume of pores)/volume of column tube}$$

This is the fraction of the column tube volume that is the elution volume of unretained marker compounds, such as D_2O in aqueous systems. In practice with porous silica, the total porosity is about 0.70 and hence a rule of thumb is that the 'solvent front' elutes in three-quarters of the column tube volume. This volume is tabulated for a few popular sizes below.

length mm	diameter inch	bore mm	volume ml	¾ volume ml
100	1/4	5	2.0	1.5
200	1/4	5	3.9	2.9
250	1/4	4.6	4.2	3.1
100	3/8	8	5.0	3.8
200	3/8	8	10	7.5
500	1	22.5	200	150

Note that porosity is quite distinct from permeability: porosity concerns volume, usually expressed as a volume fraction; permeability is the ease with which liquid can get through the porosity.

The permeability of nanometre-sized pores is negligibly small (because flow in tubes is proportional to $1/d^4$) and molecules gain access only by diffusion.

The porosity of beds of packed solid particles used in LC (and GC) is about 0.4. This means that nearly half the column volume is full of liquid (or gas). (Note the different porosity between the maximum density packing structure for spheres of 0.26, and the least dense at 0.88.)

In adsorption chromatography, the particle porosity is important only in as much as the solid material away from the pore surface does not separate while each whole particle contributes to band broadening around itself. But much more important is the surface area that is effective at separating.

In permeation chromatography, the whole pore volume is used to separate by molecular size, so the pore volume should be as large as possible. The difference in elution volume between large (excluded) and small permeating molecules is only the pore volume of the particles and cannot be extended as in adsorption chromatography.

Beyond a porosity of ~0.9, that is only 10% of the particle is solid, the particles become impractically weak and break-up under pressure. Designers of GPC materials must try to keep the pore volume as high as possible while still creating the right hole structure.

Packing and porosity

In theory we want to pack the particles in as closely as possible so that the interstitial porosity is as small as possible, but in practice small increases in packing density are achieved at a high price of pressure requirement as the permeability will also be smaller. More loosely but chromatographically well-packed beds must be the real target. These two requirements are reflected in low values for the Separation Impedance (see section 3).

The interaction between packing methods, dry or slurry, flocculated or deflocculated, together with concentration, packing pressure and the resulting interparticle porosity and permeability has hardly been studied yet, apart from the observation of lower permeability of dry-packed compared to slurry-packed columns by Knox and coworkers[109].

Pore Size

summary
- **pores 5-10 nm most useful for adsorption chromatography**
- **little good data on effect of varying pore size on selectivity and loading capacity**

The effect of pore size, and its distribution, has been little studied to date. If the separation is to be made on the surface (adsorption chromatography), as opposed to sizing (permeation chromatography) the pore size can be as small as is consistent with fast enough transfer of molecules down the pores, the thus sharp enough peaks.

The pressure to reduce the pore size comes from the need to increase surface area. (Usually the specific surface area in m^2/g is quoted: the area per column volume is the chromatographically important factor.) The solid part of the column packing material does not separate, it only supports the surface. Nor does the interstitial volume contribute to separation. So the porosity of the particles should be high, that is have a high pore volume, and the interstitial porosity of the column as low as consistent with good packing and permeability.

Surface area and pore size are linked: beyond a certain point the area can only be increased by reduction in the pore size. Most manufacturers have chosen a compromise of 6-10 nm pore size and 200-400 m^2/g surface area. Processing methods may be one factor which influences the choice of these values. Data from Eisenbeiss[335] suggest that 4 nm pore Lichrosorb retains more strongly than 6 or 10 nm types: retention (k') is nearly proportional to surface area, but the selectivity of the smaller pore material is lower. So better separations may be obtained on larger pore particles especially if the eluent composition is adjusted (reduced in strength for low surface area materials) to make the degree of retention the same regardless of surface area.

Vermont, Deleuil, de Vries and Guillemin[331] describe work with even smaller pore materials, down to 2 nm. (These are Spherosils, some commercially available, and some have Porasil equivalents.) They conclude conversely that a very high surface area $\sim$800 m^2/g is best for selectivity.

The relative loading capacity of these materials for preparative work has yet to be determined, but again, up to a point we must expect that a high surface area will be an advantage.

Whether the problem is analytical or preparative, it may be that the best material will vary with both the size or type of solute molecule, and with the type of solvent system in use.

Assessment of the Pore Structure and Surface Area

The measurement of pore structure, both size and distribution of size, and shape, and the total surface area is a complex subject. Linsen and Broekhoff[329] have reviewed the methods and problems. The two main methods are gas adsorption and mercury porosimetry.

Gas adsorption

summary
- **useful information on surface area, pore volume, pore size and distribution, and some information on shape**
- **but nitrogen is not a typical solute**

The adsorption and desorption of gas, usually nitrogen, is recorded as a function of pressure (partial) with the sample cooled in liquid nitrogen.

From mono-layer adsorption of nitrogen at low pressures the surface area can be calculated by using the BET equation (after Brunauer, Emmett and Teller[330]).

At higher partial pressure, multi-layer adsorption takes place and eventually the pores fill by capillary condensation, allowing estimation of the pore volume. On desorption the nitrogen evaporates from a menicus at different pressures compared to adsorption thus giving a hystersis loop in the isotherm from which the distribution of pore sizes can be assessed.

From the shape of the adsorption–desorption isotherm some tentative information about the shape of the pores can be inferred. Characteristic curves are obtained for cylindrical pores, open at both ends, the desirable shape for chromatography, or so-called ink bottle pores with narrow necks and wide bodies, only open at one end. Diffusion in and out of these can be expected to be rather slower. With nitrogen some complications are that the capillarity relationship breaks down with pores below about 1 nm, and there are some spaces where the particles touch which are like pores, but whose character will depend on the packing structure as well as the pore structure.

From a chromatographic point of view, nitrogen is not a representative molecule, nor even a reasonable model for one. Other inert gases like argon, or small alkanes like neopentane may prove better models. So surface areas estimated from this method can only be expected to correlate approximately with chromatographic results. In particular the smallest pores, say less than 2 nm, may contribute much to the surface area but little to chromatographic performance. This is seen in the behaviour of materials with very high surface areas >400 m^2/g.

Mercury porosimetry

Mercury porosimetry has been widely used to measure pore size, volume and distribution. The pressure required to inject mercury into the pores is recorded as a function of volume. Unfortunately the results do not correlate completely with estimates of the pore sizes from permeation chromatography[23], and there is some evidence that the pressures can break the walls of some pores[271].

The pore diameter (nm) $\approx$ 150 000/P(bar) so about 15 000 bar is required to force mercury into pores of 10 nm diameter.

Neither of these methods is ideal from a chromatographic point of view and some chromatographic probe would be better. At present there seem to be even more complications in the use of any probe of this type and instead we must make careful use of the data available from these two methods.

Pore volume

This can be measured by the method of Fisher and Mottlau[332] or calculated from other parameters, particularly the shape of the adsorption–desorption isotherm or mercury porosimetry.

The relationship between pore volume, pore size and surface area

These parameters are interlinked. A theoretical model with cylindrical pores predicts (from geometry) the relationship[323,99]

$$\text{pore diameter (nm)} \approx 4\,000 \times \text{pore volume (ml/g)/ specific surface area (m}^2\text{/g)}$$

In reality, the pores in silica are irregular since they are formed from the interstices left by the aggregation of the colloidal particles in the sol from which the gel is formed. For silica these are roughly spherical, but for other materials they may be rods or cubes. So those pores which are not very close to the surface will be accessible only through other pores via rather narrower slits.

Nevertheless the above equation provides a useful guide to the practical properties of adsorbents. The pore volume is limited by the need for structural strength, which means that a high surface area can only be achieved by reducing the pore diameter.

Spherical versus Angular particles

There is considerable discussion among chromatographers about whether spherical or angular particles are best. Spherical particles will tend to yield a column of higher permeability and more regular packing structure than angular ones. It is difficult to make a fair comparison which really compares *shape alone* because available materials also vary in particle size distribution. The observed facts are that columns packed with materials which have the same mean particle size give very similar plate heights, but a higher pressure is needed with the angular particles[109].

If this mean particle size is taken to calculate the reduced plate height, then the two shapes have the same efficiency. But if the particle size is estimated from the permeability (see section 12.8) then the reduced plate height of angular particles is rather higher. The concept of separation impedance (see section 3) provides a basis for comparison which avoids this difficulty.

So there is an intuitive preference for spherical particles, but until recently their premium price tag has counter-balanced this preference. This price structure is changing. As the scale of manufacture increases, the cost of special machinery and research needed to make spherical particles the right size to start with becomes proportionately less, and the product can be sold profitably at a lower price. In the long term it seems likely that spheres will be the standard shape.

It is reasonable to suppose that it should be easier to make a more reproducible product with less batch to batch variations in particle and pore size with the processes used to make the various types of spherical particle. Each particle is derived from the same environment and each sphere receives identical treatment during manufacture, and so should have identical microstructural properties[99]. How important this is in practice is not yet clear.

Carbon

summary • **no practical materials yet for LC**
—but still a possibility for the future

Chromatographers (and manufacturers of packing materials) are busy establishing all those very many separations which are, or will be, possible on polar packings like silica and alumina. Some separations require a non-polar packing and some of the bonded phases go some way to meeting this requirement, but are not ideal, as noted elsewhere. Some form of carbon seems a natural material. Some progress and problems are reviewed by Colin, Eon and Guiochon[137]. Despite the promise of their results no practical material for the practising chromatographer is available.

Carbon on Silica

Another approach is to coat a silica support with a carbon layer[391]. This is subject to the same problems as bonded-phases with the polar surface becoming accessible to solutes through the non-polar layer.

Silica as an Adsorbent –
separation of topical steriods, preservatives, and an internal standard

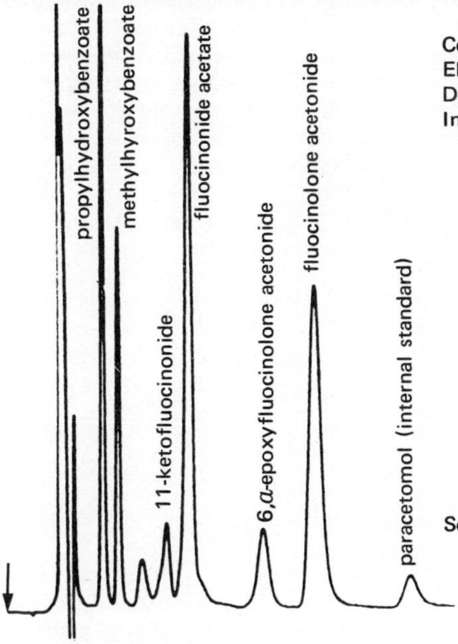

Column: Partisil 5, 100 x 5 mm bore
Eluent: hexane:ethanol:acetic acid (90:10:0.1)
Detector: CE 212 at 238 nm 0.1A
Injection: syringe on-column 2 ul

Source: R J Mellor, ICI Pharmaceuticals Division

Non-polar bonded alkyl polymer

Rapid elution of more
polar impurities and
breakdown products

Sample: 5 μl ICI PP413 (15 mg/ml)
Column: 2.1 mm x 1 m ODS-Permaphase
Eluent: 1,4-dioxan:water 5:95
Temperature: 60°C. Pressure: 50 bar
Detection: UV 254 nm 0.04A

Source: D A Warwick, ICI Plant Protection Division

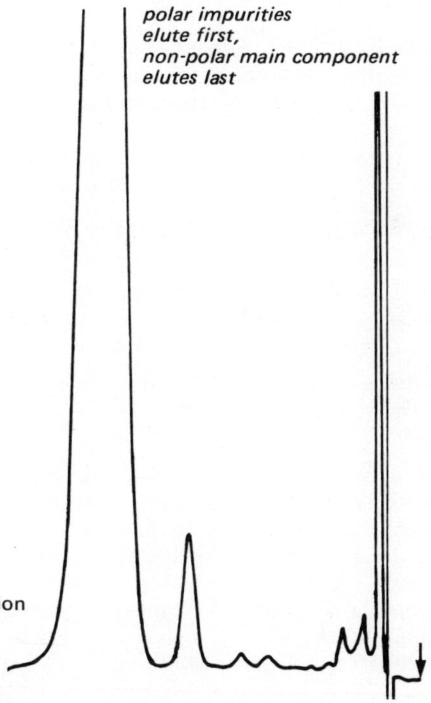

6.2 Bonded-layer materials

summary ● more sophisticated, selective, expensive and problematic than silica

The instincts of chromatographers have been to use liquids adsorbed or chemically bonded to an inert support. Historically much pre-HPLC used LLC, and many workers developing the new HPLC had worked with gas chromatography where the idea of solute molecules even getting a glimpse of the surface was anathema. As interest in LLC has waned, interest in bonded phases has replaced it.

In practice the results have proved problematical so far. Early materials have an unenviable reputation for the ease of their conversion to a non-bonded phase. Better materials are now available, and widely used, but their preparation, reliably and reproducibly, has caused some headaches for manufacturers, academics—and users. Halasz[353], with the voice of experience (perhaps bitter), characteristically sums up thus: *"It is not difficult to prepare a RP (bonded-phase), but it is extremely difficult to 'cook' a good one."*

The advantages proposed for bonded phases are:

- mechanical, thermal and solvolytic stability
- insolubility in solvents (though the bonded layer may be removed by reaction)
- rapid equilibration with the eluent
- compatible with rapid gradient elution
- sites which can interact selectively.

summary ● polymerised silane on pelliculars different from mono-layers on silica
● ODS most popular and useful

Mercifully the wild proliferation of packing phases used in GC has not yet taken place in LC bonded-phases, although potentially there are more possibilities. The handful of basic types are produced by most manufacturers: Majors has reviewed the commercially available types in 1975[214].

By far the most popular use of the bonded-phase concept has been to attach alkylsilanes, usually octadecyl, to make a non-polar surface for use in so-called reverse-phase systems (where the stationary phase or zone is non-polar and the mobile eluent is polar, the reverse of normal when the stationary silica is polar). A variety of other groups have been made but none has yet proved as popular: this may not reflect less good properties but simply that ODS is usually tried first and often works, in which case other materials are unlikely to be tried. In some ways more interesting and promising are bonded materials which have a selective group, typically an ion-exchanger, in that they have a polar surface which is quite different from that of silica. (In practice the non-polar organic backbone may have a significant chromatographic effect too.)

Rehak and Smolkova[286] give a comprehensive review of work done before 1976, updating an earlier review by Pryde[38], and complementing the commercial materials review by Majors[214].

The original concept was that all surface silanols should be reacted to give a mono-molecular layer of the organic species. During realisation of this concept some complications have become apparent:

- new silanol groups can be made by hydrolysis of siloxane groups on the surface.
- not all silanols will, or can, react, partly for steric reasons.
- eluents may remove some bonded groups, for example by hydrolysis.
- eluents can dissolve the silica core at high pH.
- reagents may polymerise and produce a thicker coating. This phenomenon is especially prevalent with di- and tri-chlorosilanes which cross-link with any water molecules present.

Bonding methods

summary
- **≡Si–O–Si–C most common link**
- **≡Si–C more difficult to prepare**

Because it is near the surface and shielded from eluent and solutes, the gluing group has little effect on the chromatographic performance. But it has a critical effect on the practical usefulness of a material because the highly polar surface is exposed if the non-polar group is lost and this changes the chromatographic properties drastically.

The silanol–chlorosilane bonding route has been most widely adopted, particularly by manufacturers to whom the availability and cost of reagents is important.

Grignard reagents like benzyllithium have also been used but less frequently because they are troublesome to prepare. Saunders and co-workers[249] chose this method because it could be modified by a wider range of reagents than Si–O–Si–C based materials because the Si–C bond is more stable, especially towards electrophilic reagents.

Bonded materials

Octadecyl, and other alkyl chain length, siloxane-bonded materials are available from nearly all manufacturers. As discussed below, these vary in the degree to which they are polymerised. The behaviour of mono-layer ODS–silica is quite different from pellicular polymerised ODS. This may be because of a higher organic content of a given length of column, or because silanol groups affect the separation, or because the mechanism of separation is different in that the liquid inside the pores is altered by the organic non-polar layer on the walls, a phenomenon mentioned again below. In practice this different character may, or may not, be helpful. It is certainly a barrier to changing from methods based on inefficient pellicular packings to much more efficient microparticle silane-on-silica columns.

Typical recipes for making ODS–silica from silica and octadecylsilane are given by Kingston and Gerhardt[36] and Wheals[356].

In order to produce more complex groups like ion-exchangers, it may be better to attach a backbone organic group to the silica, and then add the functional group.

Vinyl[362] and γ-glycidoxypropyl groups[360] are typical examples and lend themselves to making a very wide range of possible materials. (However the need for such a wide range needs to be established before too many are put into production.)

Some common groups are:

$-CH_3$	methyl	$-(CH_2)_n Cl$	chloroalkyl (for reaction with amines to give ion-exchanger)
$-(CH_2)_n CH_3$	alkyl (C_4-C_{18})		
$-CH=CH_2$	vinyl		
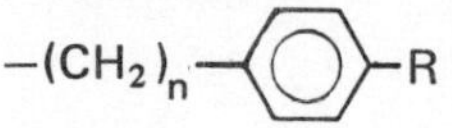	(for further reaction)	$-(CH_2)_n CN$	cyano
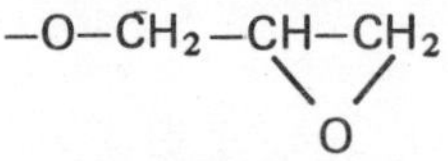	oxirane (plus further reaction)	$-(CH_2)_n NH_2$	amino
		$-CH_2SH$	thiol

The most selective phase so far described[361] (apart from the affinity chromatography phases which are specific for one molecular type) is a phase where the bonded organic group terminates with a tetranitrofluorenimino group, known to be capable of charge transfer. This gives high selectivity for polynuclear aromatic hydrocarbons, but too broad peaks to be a practical alternative to very high efficiency silica columns (see section 8.11).

Alkyl chain length

Halasz[353], who has reviewed the preparation and properties of reversed phases, concluded that while a C_4 alkyl chain length gave optimum coverage judged by the capacity factor (k') of nitrobenzene (C_1 0.34; C_4 0.11; C_{18} 0.47) with *n*-heptane as eluent, when used in the normal way with polar eluents like methanol—water it was better in practice to have the long retention resulting from the longer chain of C_{18} and hence a 'thicker' layer.

Other workers such as Kirkland[365], Majors and Hopper[364], and Hemetsberger, Maasfeld and Ricken[363] agree with this general conclusion that a long chain length is best. Hemetsberger makes two important practical points. The more retention a phase provides, the lower concentration of water is needed for a given k'. Since methanol has a lower viscosity (0.6 mN s/m^2), and higher diffusivity, than water (1.0) or even more important, the annoyingly anomalous high viscosity of alcohol—water mixtures (see section 12.6), the practical result is several-fold lower column pressure drops, and appreciably narrow peaks because the diffusivity of solutes is higher[363]. For preparative systems, solubility is important and most solutes which might be separated on this alkyl-bonded phase are much more soluble in methanol than water.

Knox and Pryde[237] however argue that in practice a short-alkyl chain sometimes gives better separations, for reasons discussed below.

Pore size

summary
- **optimum size not certain**
- **octadecyl group is ≈ 1 nm long**

The optimum pore size is not yet certain. Although ODS has been bonded to some silicas with mean pore diameter ~5 nm (and thus some pores much less than 5 nm diameter), Halasz[353] observes that the total porosity of these materials suggests that so much of the pore is full of ODS that rather little is available to the eluent and the rate of equilibration may be reduced compared to a larger pore.

Knox and Pryde[237] using 10 nm pore silica, prepared materials which had efficiences the equal or better of the silica before any organic material has been added. This suggests that a larger pore size is better judged by the efficiency of criterion. On the other hand the amount of retention, or capacity, of a material with rather full pores may sometimes be preferable. Commercial materials of both types are available.

Silica surface for silanization

summary
- **surface should be fully hydroxylated**
- **but free of physisorbed water, unless a cross-linked layer is wanted**
- **not enough room for all hydroxyls to bond to long alkyl chains**

The surface of the silica is best prepared by maximizing the number of hydroxyl groups on the surface, usually by boiling with water, or acid, and the drying at ~150°C, perhaps in a vacuum. If the silica is heated above this temperature, the number of silanols will be reduced by dehydration rather than drying off physisorbed water. If the surface is not fully hydroxylated to start with, new silanols may be formed by hydrolysis during use with aqueous eluents.

In the fully hydroxylated state most determinations have put the concentration of silanols at ~5 OH/nm^2. Halasz and co-workers[353] managed to react nearly all these hydroxyls with a C_1 chain length silane, and thus showed that they needed only 0.2 nm^2 per group, but reacted less than half the total hydroxyl groups with C_{18} groups, which needed about 0.5 nm^2. As mentioned above the overall conclusion was that a longer chain had better separation characteristics. There is not enough room for all silanols to be reacted with an organic group of this chain length, so some silanols become shielded and remain unreacted. In theory these groups should also be shielded from solute molecules, but this is less certain in practice.

Residual silanol groups

summary • **if few silanols**
- **no methyl red adsorption**
- **$k' < 0.5$ for nitrobenzene with alkane eluent**

The number of 'visible' silanol groups can be assessed from measurements of retention with solutes like nitrobenzene using a pure alkane as eluent. For 'good' materials k' should be <0.5. A simpler, but less sensitive, method is to test for a red colour caused by methyl red adsorption[357]. Adsorption of cytochrome c has been used to assess the coverage of a glycophase bonded-layer[269] intended for use with sensitive biological molecules.

Although materials which have more silanols available for interaction with the solute are quite usable, it seems sensible if one is seeking a RP material, to use this as a criterion of good quality. It will be difficult to manufacture material of reproducible performance if there is a significant silanol activity from a non-polar bonded phase material.

Hemetsberger and co-workers[363] and Majors and Hopper[364] deduced from the IR spectrum that silanol groups are still present even after the reaction with long alkylchlorosilanes is complete, and yet these groups cannot react with trimethylchlorosilane (TCMS). Kirkland[365] also failed to obtain any significant reaction after bonding ODS. (But with a phenyl bonded phase it was possible to see methyl groups in the IR spectrum after bonding[363].)

Mono-layer or polymerised layer?

summary
- **thicker polymerised layer easily formed with traces of water**
- **thick layer has higher capacity, retains more—but is less efficient**

Thicker layers of alkylsilanes can easily, and often inadvertently, be produced if water is present to cross-link the di- or tri-functional chlorosilane.

Gilpin, Camillo and Janicki[359] show how increasing the level of water physisorbed on the silica surface increases both the degree of retention and the plate height. The latter is caused by slower equilibration with a thicker polymerised layer. They controlled the water level by first saturating with wet toluene and then passing different volumes of dry toluene through to remove different amounts of the adsorbed water.

In practice it is difficult to exclude all water from the reaction, and unless rigorous precautions, like distilling the silane directly onto the silica under a dry atmosphere, are taken, some cross-linking will occur.

Thicker layers have the disadvantage that mass transfer is slower, but on the other hand if the amount of the organic material is higher for the same sized column, there is less chance of silanols being visible to solute molecules, and removal of the polymer layer is more difficult because the attaching links are less accessible and more numerous. If the layer is thick enough, the column impedance may rise slightly[356].

So provided very small particles $\sim 5\ \mu m$ are used, the extra capacity and robustness may be the most effective compromise as long as, say, the increased C-term does not lead to more than a doubling of plate height at useful speeds, say for $\nu \simeq 5$.

Separation mechanisms with bonded materials

summary
- **separation mechanisms still not well understood**
- **simple reverse-phase concept not sufficient**
- **non-polar solutes separate by a pseudo-partition mechanism and long alkyl chains give more retention**

Two types of mechanisms have been the subject of discussion (the prerogative of authors preparing a bonded phase). As with silica itself, the mechanisms proposed are a function of both the solute and the eluent.

The simplest approach is to visualise a bonded-phase as a captive liquid, with properties similar to octadecane for octadecylsilane, for example. Locke[366] has explored this concept and his data, and that of others supports this hypothesis. Hemetsberger[363] however gives an example of how shape, linked to the entropy contribution and the difference in free energy of distribution of two components, has an important effect on selectivity.

So separations on bonded-phases like ODS with polar eluents are analogous, but with reversed polarity, to separations on silica when used with non-polar eluents like hexane, without added polar components like methanol which interact very strongly with the surface. A linear relationship between bonded alkyl chain length and retention proves that a common mechanism applies, for example for bromobenzenes on alkylsilanes[363].

But, like silica, if the eluent contains components which are like the surface in polarity, for example methanol in hexane for silica, or in water for alkylsilane—silica, then a layer rich in this component may form near the surface and provide a different basis for separation. Knox has coined the term 'eluent polarisation chromatography' for this phenomenon.

summary
- **polar solutes separate on the polarized eluent surface layer hence for these fairly short alkyl chains may be better**

Knox and Pryde[237] and Kikta and Grushka[306] discuss how the eluent composition in the pores is altered by the bonded phase. In an organic solvent—water system the less-polar organic solvent will solvate the bonded alkyl layer creating a zone in the pore which is organic rich. Partition of solutes between this and the bulk eluent may be the main separation mechanism but work leading to a definitive conclusion is still awaited. Separation of tetracyclines[119] which are highly polar and insoluble in alkanes on a short-chain alkyl—silica is certainly difficult to account for in conventional reverse-phase terms and this mechanism is much more plausible. Knox and Pryde observe *"The short-chain alkyl-bonded silica is evidently much more versatile than would at first sight be expected. But unlike conventional adsorbents such as silica or alumina the SAS—silica preferentially extracts the non-polar component(s) rather than the polar components of an eluent. It is thus the true reversed phase analog of a normal adsorbent such as silica gel or alumina."*

Eluent polarisation

Si—OH---HO—Me---alkane	Si—O—$(CH_2)_n$—Me---Me—OH---water
normal phase	**reverse phase**

The effect of Bonded–ODS Loading on silica

Column: 250 mm x 5 mm bore
Packing: octadecylsilane(ODS) on Partisil 5

Source: B B Wheals[356]

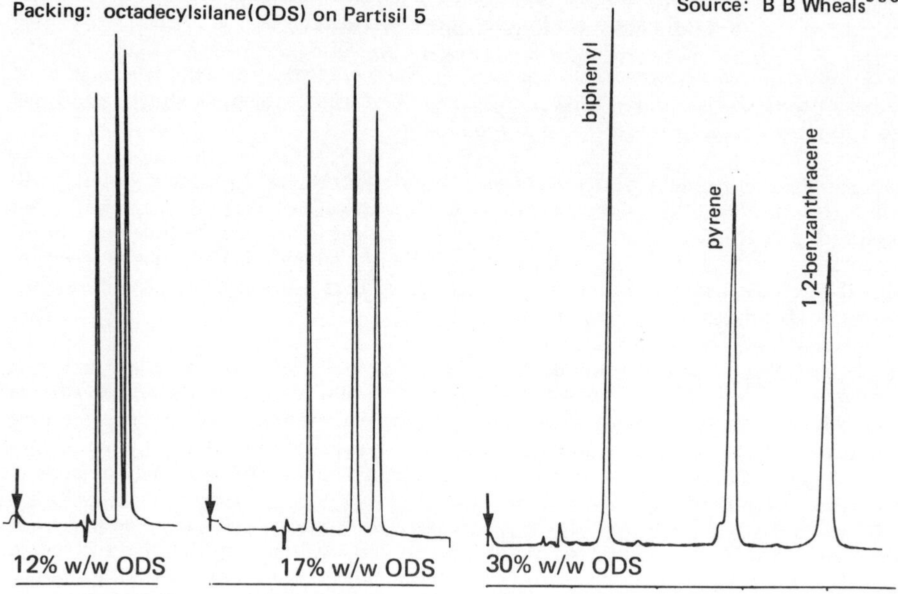

Packing reverse-phase materials

summary
- **balanced density halocarbon slurries are deflocculated, but expensive, toxic, viscous and difficult to remove**
- **acetone also deflocculates and has very low viscosity**

The general methods and problems of packing are described in section 4.2. Balanced density methods are widely used and may give a deflocculated but viscous slurry. Acetone is just non-polar enough to deflocculate even alkyl-bonded materials, and has the advantages of cheapness, low toxicity and very low viscosity. Cox and co-workers[247] report that it can also be used successfully with polar materials like a bonded ion-exchanger and silica itself.

6.2.1 Liquid–liquid chromatography

summary
- **no longer popular except**
  - **ternary component – two phase systems**
  - **for determination of partition coefficients**
- **viscous stationary phases cause too much peak broadening**

Unlike GC, where nearly all separations are achieved by gas–liquid chromatography, the analogous process in LC, liquid–liquid chromatography (LLC) has never proved as successful or popular. Many early workers in LC extrapolated from their experience in GC and assumed that LLC would be the most useful separation mode, but adsorption chromatography liquid–solid chromatography (LSC) has at least *appeared* to be most useful in LC. Stewart and Perry[350] saw the problem clearly in 1968, *" . . . this is because immiscible phase pairs tend to differ very considerably in polarity (e.g. water–hexane) and sample components tend to have partition coefficients in such a system tending either to zero or infinity."* (And they proceed to describe bonding an octadecylsilane layer.) In practice, and theory, the distinction between LLC and LSC, or partition and adsorption, is by no means as clear-cut as implied by the terminology – and most texts. The distinction between stationary and mobile phases is also much less clear than with GC. It is perhaps more helpful to consider two zones. The mobile zone is the liquid flowing outside the particles which contains solute molecules. The stationary zone is the environment inside the pores or structure or surface of the particles in which solute molecules are retained. Whereas the mobile zone can be assumed to be much the same as the eluent being pumped through the column, the environment in which separation takes place is much more difficult to determine and describe. Some examples of the problem are layers of polar liquids like water on silica, salting-out phenomena, the role of solvent in a swollen organic resin, and the separation of non-ionic compounds on ion-exchanging materials.

It is fortunate indeed that it is not necessary to *understand* in order to *separate*.

High-efficiency LLC

summary
- **ternary liquid–two phase systems give good efficiency and selectivity**
- **but can be troublesome in practice**

Because viscous liquids cause so much peak broadening under high speed LC conditions, several workers have turned to systems using non-viscous solvents[347,348,349,354,355]. Since the components diffuse in and out of the pores quickly an equilibrium can be achieved reasonably quickly. Parris[355] suggests that 25-50 column volumes are required. Solute molecules equilibrate with the stationary pore liquid at a similarly high speed and thus cause little zone broadening. Some systems that have been used are:

water–ethanol–2,2,4-trimethylpentane[348]
water–isopropanol–dichloromethane[347]

The distinction between the actual stationary liquid in these systems and those in the pores of silicas used in separations which are ostensibly adsorption chromatography is blurred. A new conceptual framework may result from an improvement of our understanding of why separations work.

Why choose LLC?

Although LLC has gone out of vogue there are *some* good reasons to preferring it to LSC.

- Direct transfer of classical LLC methods is often feasible. The efficiency should rise, but retention volumes and column capacity ratios (k') stay the same.

- The process is equivalent to counter-current distribution and batch methods of measuring partition coefficients (see below), and results from these techniques should correlate with LC. Huber and co-workers[348,354] have shown that with painstaking attention to detail, a precise correlation of dynamic and static methods can be obtained for steroids in the system water–ethanol–2,2,4-trimethylpentane.

- It is gentle in as much as there can be no reaction on the surface. In practice this is a surprisingly rare problem. In general, if the compound survives solution, it will survive chromatography.

- Selectivity can be usefully different from adsorption systems. Gonnet and Rocca[347] claim " . . . *une bonne selectivitie* . . . " as an important reason for using ternary solvent systems.

There are, of course, some disadvantages:

- The high viscosity of some liquids, and polymers, means that it is almost essential to operate above ambient temperature. This increases the cost of the equipment and is much less convenient in operation. Problems of ovens are discussed in more detail in section 7.1. This problem does not arise with low viscosity liquids, so it is more a question of selecting the right liquid.

- The solvent for the sample to be injected must not be too different from the eluent or it will displace the stationary liquid, and after many injections may become a new stationary phase component, or precipitate the sample. The former will slowly change retention volumes in an unpredictable way; the latter will lead to disastrous tailing.

Support materials for LLC

summary
- **large pores, good pore volume, but low surface area needed**
- **diatomaceous materials suitable except for particle size and shape**

Materials used to support stationary liquid phases should have quite large pores, and a small surface area, and ideally a high pore volume. High surface area materials can be used but there is a danger that adsorption mechanisms will also play some part, which may be undesirable leading to poor peak shape or variations in retention times. This requirement is similar to that for GC where adsorption is undesirable, and dealt with by silanisation if it proves troublesome.

Pellicular materials are usable, though their capacity is rather low and small particles with a large pore volume may be better. Until recently suitable microparticles were not available and so much LLC work has been done using pelliculars.

A *non-polar surface* may be needed to bond a non-polar stationary phase. The normal method of meeting this need is to silanise the surface: both short alkyl groups like trimethyl silane, and long ones like octadecylsilane have been utilised. An easy way, in practice, is to start with a phase bonded onto a pellicular or porous material.

Diatomaceous earths as used in GC have very high pore volumes (1-2 ml/g) with ~1 μm pores and a low surface area of 1-5 m^2/g (ref. 9, p 28), but are difficult to obtain and size small enough for HPLC, so column efficiencies are low.

Coating the support with the stationary phase

Polymerised layers will normally be added by the manufacturer and will not be discussed further.

Several methods of coating liquids which have been used are:

Direct. The liquid is added directly to the support and tumbled, and/or heated. The term 'toasted' has been used to describe a particularly tightly held liquid β,β-oxydiproprionitrile (BOP) on a very high surface area porous silica[331].

Solvent evaporation. A more even distribution can be obtained by dissolving the viscous liquid in a mobile solvent, for example BOP in dichloromethane. After coating, the solvent is evaporated off. The process is widely used in preparing GC column packings. Because the solvent is often more polar than the liquid to be coated it is often strongly retained by the support, especially if it has a high surface area, and it may be much more difficult to remove than in the high temperature environment that most GC columns operate in.

Precipitation. The stationary liquid can be precipitated out in the pores. A solution of the liquid is pumped into the column, and then a second immiscible liquid pumped through to sweep away the diluent leaving the stationary phase in the pores.

In situ. The most convenient method is to pump the liquid onto the column, either as a component in a two of three component system and allow it to reach equilibrium, or to inject slugs of the stationary liquid, perhaps diluted with some eluent so that it coats the column. It is of course essential to get the system into equilibrium, which may be a slow process. The main advantage of this method is that the column can be reconditioned by injecting further slugs of stationary liquid as necessary. A true equilibrium system should never need rejuvenation. In practice the problem is that components of the injected sample become irreversibly (under the chromatographic conditions in use) onto the support and change the separation, or even block the column.

Pre-columns

These are widely used with LLC systems to avoid disturbing the column equilibrium too much. It may be desirable to use a pre-column for the sample feed as well as that commonly used in the eluent feed line. The pre-columns can be replaced when contaminated. In practice, now that slurry-packing is established as a very simple and quick process, repacking the whole column may be nearly as easy.

6.2.2 (Bio)Affinity Chromatography

summary
- **extremely selective bonded phases**
- **important for separating biological macro-molecules**

Outside the mainstream of LC development is affinity chromatography, a most important separation technique for complex biological molecules. A highly specific molecule, typically an enzyme inhibitor or co-factor, is bound (bonded, immobilized or insolubilized) to a gel. If a sample is injected onto the column, the enzyme will be selectively retained by the bound inhibitor molecules. A fairly typical example is described by Amneus, Gabel and Kasche[282] where soybean trypsin inhibitor attached to Sepharose is used to fractionate trypsin enzyme. The bonded molecules can be considered as a very carefully tailored analog of the alkylsilanes used to separate much smaller molecules.

Like the bonding of smaller molecules, such as alkylsilanes on silicas, there are many practical problems and the mechanism of separation is rarely clearcut. The matrix is usually a soft large-pore gel like agarose: very large pores are essential to allow the large molecules to permeate and reach the bonding molecules. As with materials used in HPLC, the matrix inevitably plays some part in the separation process, either by sizing or adsorption, polar or non-polar.

Bonding may be via a chemical link like that used to bond alkylsilanes used in HPLC, but the problems of attaching large molecules with a bifunctional reagent to a soft gel are even greater than those with small molecules and covalent coupling is often preferred. Because the reagent must be bifunctional, some does not link to a specific molecule and leads to non-specific adsorption. It is important, but not easy in practice, to immobilize only pure material. The attachment process can also alter the properties of immobilized molecules, and hence their affinity for the molecules to be separated.

For all these theoretical complications and practical difficulties, the technique is yielding many useful separations[25]. A useful review has been given by Williams and Smith[281]. The increasing number of applications are listed and reviewed briefly by Weetall[212] in a periodic technique review in Analytical Chemistry. J. Porath[390] discusses critically the methodology and application of what he now proposes should be called "bioaffinity" chromatography to describe better the *biological recognition* phenomena.

6.3 Ion-Exchange Chromatography

summary
- **separation mechanisms complex and often only partly ion-exchange**
- **traditional resins very useful but not ideal**
- **pellicular materials faster but very low capacity**
- **mono-layer bonded phases on silica still being developed**

This branch of liquid chromatography developed to give 'high performance' many years before developments in adsorption chromatography made HPLC fashionable. The processes – and the limitations – were well understood from the work of Hamilton[105] and the search for higher efficiency led to the use of slurry-packed microparticles[233,234] long before these came into vogue for silica. Indeed some of the impetus for development of LC equipment came from progress with ion-exchange materials, especially pellicular materials for nucleic acid analysis (see section 2). And of course ion-exchange has long been used as a preparative procedure.

The subject is extensively treated elsewhere: three texts[15,16,22] are chosen for listing in the bibliography for their analytical emphasis: many others are available. C.D. Scott[243] has reviewed the application of high pressure ion-exchange chromatography (IEC) to biochemical problems. Only a brief outline will be given here with emphasis on the improvements that can and are being made, particularly to bonded ion-exchange groups on inorganic microparticle supports.

Although the simple ion-exchange principle is straightforward, the actual processes in separations on ion-exchange materials are usually complex. There are several types of separation mechanism, and often more than one is operating. All organic ion-exchange materials, whether porous resins or mono-molecular layers on silica, are also capable of partition chromatography as if they had no ionic sites. Inorganic materials, typically silica, used as a support may also act as polar adsorbents. All materials can separate by size. And the eluent may, as with other forms of chromatography, alter the solute by protonation, deprotonation, solvation or ion-pair formation. So predicting the effect of changing conditions may not be easy, as expressed by Horvath and Lipsky[254] " . . . *this complex nature of ion-exchange chromatography, which is invaluable for achieving the high selectivity characteristic for this technique, also makes data interpretation and prediction of solute behaviour difficult, and optimum separating conditions must often be found by trial and error.*"

Ion-exchange

Separation takes place in IEC because of competition between ions in the eluent and the solute for oppositely charged sites on the stationary resin.

Thus for example with a cation exchanger

solute cations X^+ compete with eluent cations C^+ for sites like $R\text{-}SO_3^-$

and similarly

solute anions Y^- compete with eluent anions A^- for sites like R_4N^+

If the solute bonds strongly to the resin it will be well retained whereas if it only interacts weakly it will be displaced by the eluent ions and elute quickly.

The strength of the eluent ion's competition for the exchanger site also controls the retention of solute. Tabulations of the relative strengths of ions are given in most texts, and for a few ions below. Divalent and large ions tend to bond most strongly to the exchanger, but in practice the situation is more complex.

For anion exchangers

$$F^- < \text{acetate} < \text{formate} < Cl^- < Br^- < I^- < NO_3^- < SO_4^=$$

and for cation exchangers

$$Li^+ < H^+ < Na^+ < K^+ < Rb^+ < Ag^+$$

The ion-exchanger tends to prefer the ion with

- high valence
- small (solvated) size
- high polarisability
- strong interactions with either ionic groups or matrix
- weak interactions with solvent

Modification of solute state by the eluent

summary • **protonation, deprotonation or ion-pair formation common**

The solute is rarely chromatographed in the form in which it is injected. The eluent, especially the pH, controls the state and concentration of solute ions, and ion pairs. Acids will be protonated at low pH and thus only small concentrations of A^+ will be able to compete for ion exchanger sites. Thus the acid HA will be eluted quicker as the pH is reduced, unless the free acid is non-polar enough to be retained by a reverse-phase mechanism on the resin matrix. Bases will be protonated at low pH and higher concentrations of BH^+ compete for the exchanger sites and thus be retained.

So acid conditions lead to elution of acid solutes, and
basic conditions lead to elution of basic solutes.

Some fine tuning of the pH allows control of the elution order and resolution of amino-acids, an important analytical problem. Hamilton describes the effect on acidic, neutral and basic amino-acids in a classic study[252].

Sugars are often chromatographed as borate anion complexes which will anion exchange in sodium tetraborate-boric acid buffer whose ionic strength is increased from 0.2 to 0.8 M[266]. Separation of sugars by other eluents does not involve ion-exchange.

Supports for Ion-exchange sites

summary
- **organic resin (polystyrene) matrix popular but disadvantages of low diffusion rate and swelling**
- **silica or alumina matrix rigid**
- **macromolecules need large pore matrix like cellulose**

The most popular support for the ion-exchanging sites is a styrenedivinylbenzene copolymer.

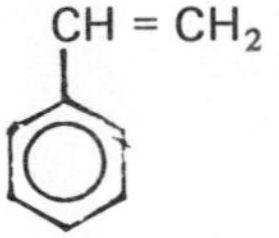

styrene
polymer unit

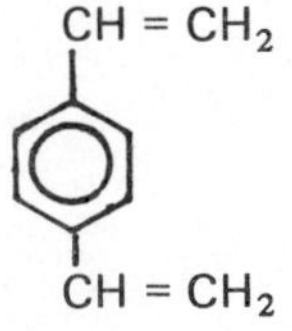

divinylbenzene
cross-linking unit

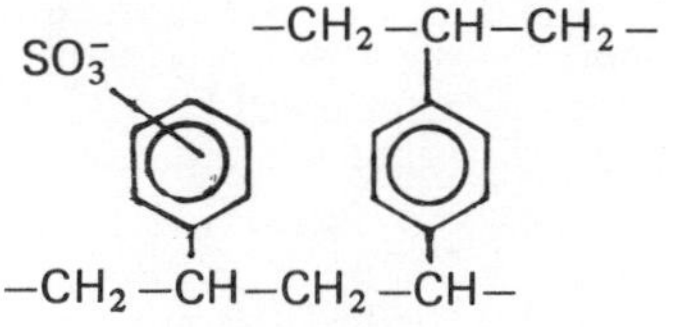

co-polymer
after sulphonation

After polymerisation, sulphonation introduces cation exchanging sites $-SO_3^-$ to produce a strong cation exchange resin (SCX).

Alternatively chloromethylation and linking with trimethylamine gives anion exchanging sites

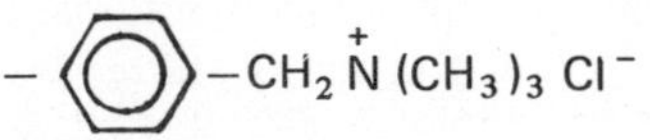

and a strong anion exchange resin (SAX).

Popular weak acid and base groups are $R{-}COO^-$ and $R{-}NH(CH_3)_2^+$

The capacity of a resin is determined by the number of aromatic rings that have added functional groups. It is expressed as equivalents per dry weight, or wet volume. Porous resins have capacities of a few mequiv/g, but pelliculars may be only a hundredth, and silica-based materials a tenth of this capacity.

Cross-linking by the difunctional divinylbenzene can be 1 – 16% (w/w composition polymerisation mixture) though ~10% is usual. Cross-linking increases mechanical strength, reduces swelling in solvents and makes the resin more selective, but reduces internal permeability and rates of diffusion. Swelling is caused by an osmosis-like force as ionic groups are solvated. Inorganic materials do not swell but may dissolve at very high or low pH, and withstand radiation well.

Pore sizes are about 1–5 nm in typical styrene resins. Cross-linking decreases the pore size. In a macro-reticular resin there is also a network of larger pores, a few tens of nm across. Pores in silica used as a matrix for organic mono-layers of ion-exchanger are about 10 nm.

summary
- **all materials separate by size**

Thus many molecules, and all macromolecules, ionic or not, are excluded from the pores of the resin. Other ion-exchanging materials derived from a cellulose, dextrins or polyacrylamide gel allow much larger molecules to permeate and ion exchange. So all supports for ion-exchange sites separate by size, regardless of whether the resin is reacted to add functional groups or not, as in permeation chromatography.

Ion-exchanging materials

Inorganic Zr, Th, Ti, Sn oxides as ion-exchangers; silica and alumina as supports for organic ion-exchange mon-layers.

Organic styrene-divinylbenzene co-polymer (cross-linked polystyrene) with sulphonic (SCX), carboxyl (WCX), quaterary N (SAX) and amino (WAX) functional groups.

[Aminex (Biorad), Durrum, Hamilton, Beckmann, Ionex (Macherey-Nagel) in sizes from 5–25 μm]

pellicular materials (nearly all 30–50 μm)

SAX $CH_3{-}NR_3^+$ WAX $-NH_3^+$, SCX $-SO_3^-$

[Pellionex (Whatman), Vydac (Varian), Zipax (Du Pont)]

cellulose with diethylamino (DEAE) anion exchange, and carboxymethyl (CM) cation exchange.

dextrans (Sephadex) and polyacrylamide with added functional groups.

Silica (and alumina) with mono-layer bonded to silanol groups:

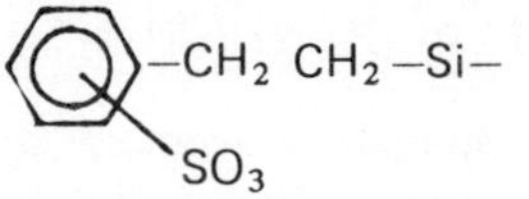

SCX

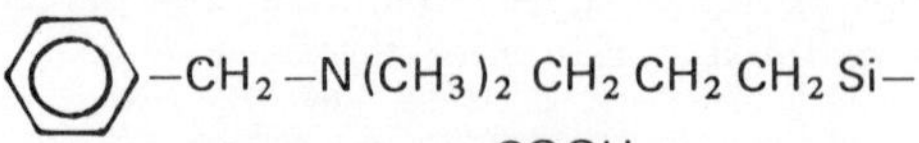

SAX

$$NH_2\,CH_2\,CH_2\,\underset{}{CH}(COOH){-}Si{-}$$

WCX

$NH_2\,(CH_2)_3\,Si{-}$ WAX

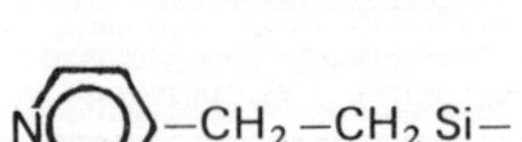

WAX

[Partisil-SAX, SCX (Whatman), Micropak (Varian), Zorbax (Du Pont]

Majors[214] tabulates in detail commercial types available in 1975.

Pellicular materials were evolved[42,43] (section 6.5) to reduce the distance the ions needed to migrate to about a micron and allowed much faster speeds, at the price of very low capacity (≈10 μequiv/g) because the column is mostly filled with rather large (≈40 μm) inert glass spheres, with only a thin layer of ion-exchanger on their outside surface. They can be dry-packed, once an advantage, now less so because slurry packing is easier and quicker.

A more recent technique, adopted because the resin which was polymerised on the surface did not always stick too permanently to the glass bead, is to bond an organo-silane to a thin porous silica layer which has been fused to the glass, and then convert this to an ion-exchanger. This may also allow one to reduce the amount of organic material relative to the ion-exchanging group and obtain a material which is more ion exchanger and less reverse-phase partition material. This may, or may not, improve separation.

The most recent approach has been to use a porous silica support (as used in adsorption chromatography) and add organic groups via the silane or other groups, and finally convert to an ion-exchanger[237,247,249]. The surface layer is now only mono-molecular, at least in theory, and diffusion rates should be as high as possible. Complicating features are the amount of organic material which can act as a reverse-phase separating material, and adsorption on residual silanol groups on any uncovered silica surface. There is no doubt that the materials give very rapid analyses at ambient temperatures[250], but manufacturing a reproducible and reliable product is said not to be easy and some problems with operating life have been reported[246].

The efficiency of columns packed with these materials have given results as good as those for silica alone[237] showing that the objectives of this development have been achieved. The capacity of most materials (≈0.5 mequiv/g) is less than that of resins but much more than pellicular materials, and their small particle size makes much faster analyses possible. There are few restrictions on solvents (except that silica starts to dissolve above ~pH 8) or pressures and flowrates.

Possible Separation Mechanisms with Ion-exchanging Materials

- size separation or permeation chromatography
- partition chromatography in reverse phase on a non-polar resin or bonded organic layer with a polar eluent
- adsorption chromatography on polar matrix, typically silanols on silica
- ion exclusion, or electrolyte exclusion
- salting-out – making eluent more polar than water by adding large amount of salt, so that less-polar components are retained by non-polar resin
- solubility – adding alcohol to aqueous eluent makes sugars more retained by a higher concentration of water in resin

- modification of solute-protonation, deprotonation or ion-pair formation

Both Cox[247], Knox and Pryde[237], and especially Asmus, Low and Novotny[253,272] discuss the extent to which separations occur by ion-exchange compared to other mechanisms on microparticle silica-based materials.

Salting out

summary • very polar salt solution causes reverse-phase type adsorption on resin

Salting out is a method of encouraging the hydrophobic (organic) solute into the hydrophobic resin by drastically increasing the salt concentration (polarity) in the eluent. The concentration of the salt solution is usually several molar, so a very soluble salt is needed. The separation of compounds (which need not be electrolytes) takes place by a reverse-phase process, that is the resin is non-polar and eluent polar. The less polar the solute the more it is retained. Sargent and Rieman[244] give a simple example of methanol, ethanol and 1-propanol. All have $k' \approx 0$ with water as eluent, but with 3M ammonium sulphate as eluent, the retention order is

1-propanol	$\gg$	ethanol	$>$	methanol
least polar—most retained				most polar—least retained

Ligand exchange

Ligand-exchange chromatography involves modifying the functional group by forming a metal—resin ion-pair, typically with Cu^{2+}, Ni^{2+} or Ag^{+}, which acts as a new and highly selective adsorbent for nitrogen-base groups (ligands) like NH_4^+, amines, amino-acids or amphetamines[245]. The eluent must compete with the solute as a ligand to the metal atom. (Although an ion-exchange resin is used, it is not necessarily ion-exchange chromatography.)

Ion exclusion

For uncharged molecules, large molecules are excluded but small molecules can permeate and are thus somewhat retained. For ions the situation is different. If, for example, NaCl is passed down a cation exchange column which is in the Na form, then Na^+ ions exchange but there is no nett change. Therefore the Na^+ ions *appear* to be excluded. Only radio-labelling the appropriate ion would demonstrate the actual exchange process. (An analogous process takes place with Cl^- ions passed down an anion exchange column in the Cl form.)

Electrolyte exclusion

If the eluent is water then the electrolyte NaCl will appear to be excluded and emerge at the interstitial volume of the bed with $K = 0$ as if it were a macromolecule too large to enter the pores. A non-electrolyte like ethylene glycol would not necessarily be adsorbed but would permeate the pores and would emerge later with a retention volume greater by the pore volume of the resin bed.

Solvent Composition within the Pores

summary • selective adsorption of polar or non-polar components within the pores can influence adsorption of non-electrolytes

Ruckert and Samuelson[255] have shown how with an ethanol—water eluent the ionic groups on the resin selectively adsorb water thus modifying the eluent composition in the pores, or more important, on their pore surface. If the organic component of an aqueous-organic matrix has a large non-polar group which is adsorbed by the resin matrix then more *organic* solvent will be found in the pores. Salts can complicate the process further by increasing the polarity of the environment outside the pores and salting out the organic phase into the pores. Similar local solvent modification processes are familiar in polar and non-polar adsorption chromatography.

These surface layers of liquid whose composition is different from that of the bulk eluent are the key to many separations. In the case of ion-exchange resins they make possible many separations of non-electrolytes, perhaps most notably of sugars, described in a series of papers by Samuelson[256]. Ordemann and Walton[397] describe how aromatic hydrocarbons can be separated on resins which have ion-exchanging groups, and show the interesting effect that the counter-ion can exert on retention of different solutes. But prediction, and rationalisation, of separation mechanisms remains difficult.

Mass transfer rates in ion-exchange materials

summary

- **mass transfer slow in resins—high C term**
- **small particle size and above ambient temperature needed**
- **pellicular – thin layer of resin better but low capacity**
- **mono-layer on silica fastest—low C term**

It soon became clear to the users of ion-exchange resins[105] that the rate of permeation through the resin matrix to and from the actual exchanger sites was rather slow and determined the feasible speed of analysis. This is in contrast to silica gels where diffusion rates of early materials were much higher. This slow diffusion shows clearly in the C or mass transfer term in the equation describing the relationship of plate height and eluent velocity (see section 3). The value of C is very much larger than with 'good' materials.

Some factors controlling the rate of movement inside the particle are:

- size and charge of ions (and molecules)
- degree of cross-linking
- temperature—increasing temperature decreases the viscosity, mainly of the eluent
- particle size—determines the distance to the furthest site
- diffusivity of solute in eluent and resin

So small particle sizes ($\approx$10 μm) and higher temperatures ($\approx$60°) became usual (despite the need to size one's own resin by a tedious elutriation[251], now unnecessary because commercial materials are available—though at high cost). Cross-linking had to be a compromise for too little gave a weak resin that would not stand much pressure thereby reducing possible flowrate and speed, but too much reduced the diffusion rate in the resin matrix.

Adjusting retention on ion-exchangers

summary • **increasing ionic strength elutes more quickly**
• **pH controls extent of ionisation**

For separating acids on an anion-exchanger Knox and Pryde[237] showed that retention

$$k', \text{ or } (t_r - t_o) \propto 1/\text{ionic strength}$$

and a similar example is shown by Scott[383] relating the retention of nucleotides on an anion-exchanger to the ionic strength.

pH has a more complicated effect because it controls the degree of ionisation. If the pH is increased acids are more ionised and more can ion-exchange and be retained. When all the molecules are ionised, pH has no further effect (unless it starts to dissolve the silica). Until this point

$$\text{retention} \propto 1/[H^+]$$

so making the eluent more acid elutes acids more quickly.

Analogous considerations can guide the situation for cation exchange.

Ion-exchange on silica

oxytetracycline and related compounds

(SCX $-SO_3H$)

packing: 5 μm diameter, ~6 nm pore, ~400 m^2/g
made to the recipe of Cox[247]
column: 200 mm, 5 mm bore
eluent: aqueous—methanol EDTA
—phosphate buffer
flowrate: ~1 ml/min
detection: UV at 263 nm
injection: syringe off-column
chart: 5 mm/min
pressure: ~100 bar

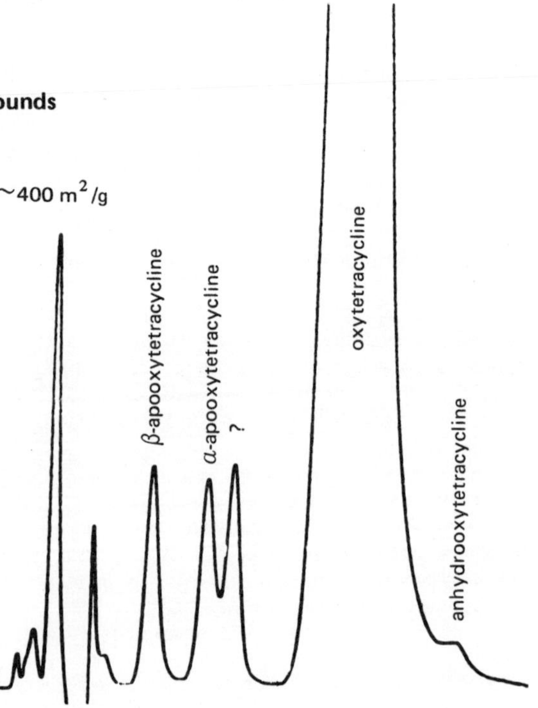

LC in practice P A Bristow hetp Macclesfield UK ISBN 0 9504833 0 3

6.4 Permeation Chromatography

summary
- **separation of molecules by size**
- **large molecules excluded and elute first**
- **small molecules which permeate pores elute later at 'solvent front' of adsorption chromatography**

Permeation chromatography[13,14] separates molecules by size. Molecules which are too big to go into the pores of the packing are excluded and elute at the volume of the spaces between the particles—the interstitial volume. Small molecules which can permeate in and out of all the pores without difficulty elute at the 'solvent front' of adsorption chromatography (provided they are not adsorbed) and have a column capacity factor $k' = 0$. Molecules of intermediate size which are a tighter fit in the pores spend more time in the flowing liquid between the particles and only rarely penetrate into the pores. Smaller molecules that are a fairly loose fit in the pores will spend most of the time in the pores and be retained more, but not as much as the very small molecules which permeate freely.

Applications

summary
- **simple separation of large and small molecules**
- **molecular weight distribution of polymers**

Permeation chromatography is applied mainly to two classes of problem:

1 Separation, usually quite simple, but often preparatively, of mixtures of big and small molecules. For example purification of biological macromolecules from low molecular weight contaminants; separation of active compounds like drugs from physiological fluids like plasma and urine (perhaps with simultaneous adsorption chromatography of the active compounds); separation of active ingredients from excipients like creams and gels in pharmaceutical formulations.

2 Separation, usually with a view to finding the molecular weight distribution, of molecules (polymers) which rarely have sufficiently different functional groups to allow separation by adsorption chromatography, but differ in size.

Preparative permeation chromatography is widely used to purify biological macromolecules. Calibration standards are also prepared this way[184], and the molecular weight of the fractions determined by other methods.

Advantages of permeation chromatography:

- narrow bands, all about the same width.
- high sample loading possible because all pore volume can be used, not just the surface (but loading reduces resolution[198]).
- short elution times—less than one column volume in adsorption chromatography.
- fewer solvent selection problems than with adsorption chromatography (but entirely different solvents tend to be used).
- retention volume can be predicted from molecular size alone (provided there are no adsorption complications).
- little column contamination from solutes or solvent impurities, because they tend to be excluded rather than adsorbed.

- Disadvantage: limited separation of similar sized molecules and limited peak capacity, usually not more than 10.

In practice there are many complications. Molecules are not a fixed size and shape. Pores are not monosize, but are tortuous and of varying length. Solvents interact with the surface and the solute. The pore surface can also have adsorption properties.

summary
- **narrow pore size distribution gives high resolution over narrow range of molecular size**
- **wide pore size distribution, as a property of the material, or preferably from mixed or coupled columns, resolves a wide range of molecular sizes**

A monosize pore structure could only separate molecules over a very narrow size range and would not be very generally useful. A simple parameter used to characterise packings is the exclusion limit (usually in the range 6–100 nm). Essentially this states the nominal size of the largest pore and thus the size of the smallest molecule that will be excluded. Real materials naturally have a pore size distribution. The effectiveness of a particular material for a particular separation depends on the precise pore size distribution. If it is wide then a wide range of molecular sizes will be somewhat excluded.

Form of Calibration curve in Permeation chromatography

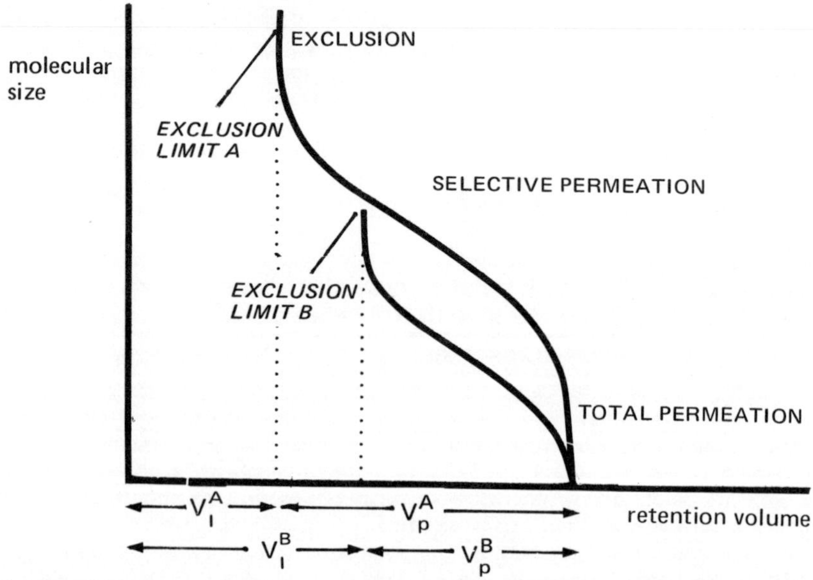

slope and curvature determine selectivity of separation with different sized molecules

LC in practice P A Bristow **hetp**, Macclesfield UK ISBN 0 9504833 0 3

The molecular size *exclusion limit* may be stated in terms of molecular weight or length. The size of the molecules which are appreciably excluded is given by the *fractionation* range. It usually covers a range of molecular sizes which differ by 10 to 100-fold. A material with a wide pore size distribution will have a wide fractional range. The fractionation range can be artificially extended by mixing packing materials of different pore size (or coupling in series columns packed with materials of different pore size). Thus molecules of almost every size can be excluded from some particles. Peak capacity is unaltered so for the same column length, the resolution of molecules of similar size is reduced.

How Permeation Chromatography differs from other forms of LC

Two factors have acted as barriers between those practising various types of permeation chromatography, and those using high performance LC. One is the hardware differences which are in some ways minor but nevertheless significant. The other is nomenclature.

Hardware requirements

Compared to adsorption chromatography permeation chromatographs need:

- high temperatures—up to 165°C, to dissolve samples and make solvents less viscous.
- lower pressures for softer packings used at present.
- valve injection—because samples are often too viscous for syringes to be used, and must be injected hot.
- less sensitive detectors, because samples are less diluted, but must be able to work at high temperatures, and a linear response to molecular weight desirable.
- accurate measurement and control of flow rate—thus siphons and constant flow pumps are common.
- recycle facility. Because samples are little diluted and restricted to one common volume, it is useful to be able to recycle[217] unresolved peaks for several passes down the column. This requires a well-swept small volume pumping system.

Nomenclature

The other factor is the nomenclature used to describe sizing chromatography. The list is absurdly long: *gel permeation* (GPC) usually of synthetic polymers, in non-aqueous polymers on fairly rigid gels like polystyrene; *gel filtration,* usually of biological macromolecules in aqueous solution with very soft gels of polycarbohydrates like Sephadex and Sepharose (filtration seems an inapt term for a system that lets large molecules through first!); *molecular sizing* chromatography; *exclusion* chromatography and *steric* chromatography; *molecular sieve* chromatography.

The generic term used here is permeation chromatography (PC) recommended by IUPAC[177] to cover all separations based on molecular size. (ASTM[211], however, in a recent publication recommend 'Liquid Exclusion Chromatography'.)

Two major reviews of the literature which have appeared recently are those of Gaylor, James and Weethall[212] in Analytical Chemistry, and who have also listed equations and definitions[274], and the Bibliography on Liquid Exclusion Chromatography by ASTM D-20.70.04.

Further confusion is caused by the different mathematical symbols which are used in a way that conflicts with usage for adsorption chromatography. This is unfortunate when many separations involve both mechanisms [23].

The diagram below tries to show the parameters of permeation chromatography alongside those of adsorption chromatography.

Distribution constant K_o is fraction of total volume of pores available to molecules of a certain size = $V_R - V_I/V_S$

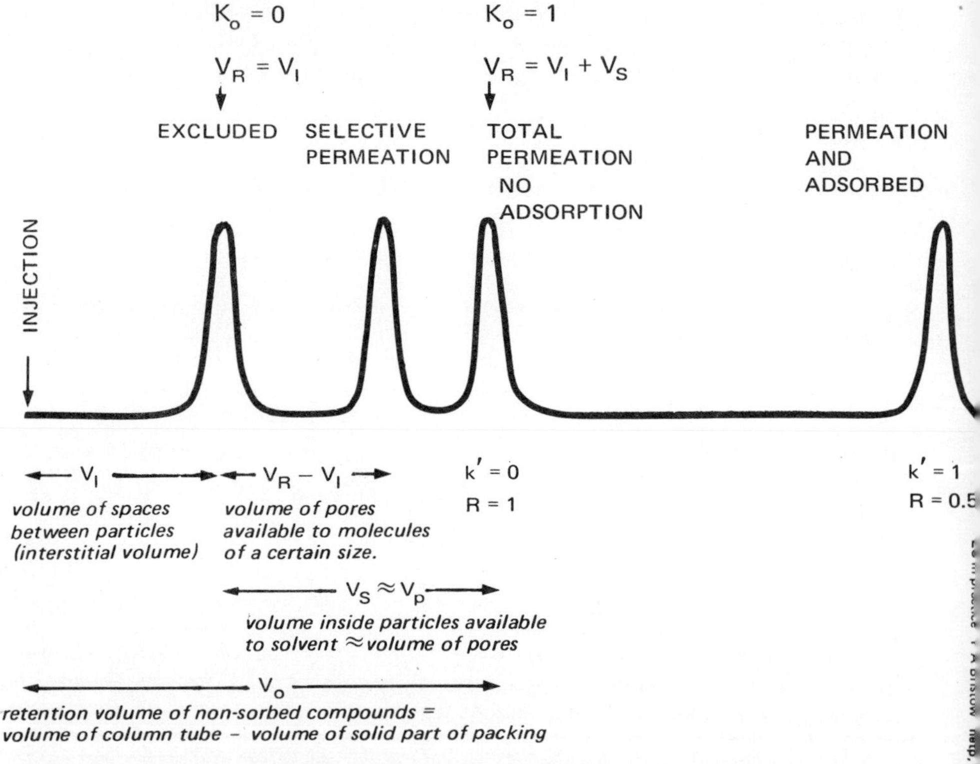

But for many molecules adsorption effects intrude on a simple size separation. It may be possible to turn this to the analyst's advantage but usually it is an unwelcome complication to an attempt to separate solely by size. Molecules which are excluded are not appreciably adsorbed for a very small proportion of the total area is on the outer surface of the particle.

Selection of the best eluent will minimise adsorption. For example the manufacturers of porous glass advise working with as high a pH as possible. Silanisation of the surface is another possible but problematic cure.

Macclesfield UK ISBN 0 9504833 0 3

The use of small particles also promises to increase both speed and peak capacity of permeation chromatography provided these problems, and the hardware problems that small particles pose, can be solved. Reported efficiencies of typical GPC column of a few thousand plates per metre are a long way from the 100,000 plates achieved in a 1 metre column[174]. A column 50 m long packed with Styragel 35 μm particles was needed to achieve this plate number. Although this comparison is unfair, some improvements are surely feasible and some progress has been made[179, 180,213,215,267]. Particularly encouraging results are reported by Kirkland[389] with specially made microspheres ~7 μm with narrow ranges of pore size, and by Scott[385] using LSC silica gel, ~5 nm pore size, to pack columns with plate numbers as high as 250 000. Majors[214] has listed some of the microparticle packings available for permeation chromatography.

It is also desirable to have a large pore volume. A particle that is almost all holes is wanted—the solid part of the particle only contributes to zone spreading. Mechanical strength limits porosity to about 0.9. Sample load capacity is also related to the pore volume of the column.

Packing Materials

summary
- **soft gels can only be used at very low pressures**
- **polystyrene widely used despite disadvantages**
- **rigid porous particles like silica have theoretical advantages but adsorption problems reduce practical usefulness**

Soft gels can only be used at very low pressures; usually the eluent is gravity fed to a glass column. Aqueous solvents pose few of the problems that bedevil designers of LC equipment. Column diameters range from about 10 mm up to 300 mm. Improvements are difficult to conceive because the polycarbohydrate gels are so soft that even a slight pressure increase will crush the gel and block the column. Nevertheless some fine separations of biological molecules have been developed and the materials are ideally developed for this type of molecule, which may itself be fragile enough to break down at high flowrates. The materials also form useful substrates for affinity chromatography which has even higher selectivity.

GPC of synthetic polymers, by contrast, requires high temperatures to dissolve many high molecular weight polymers (and to reduce viscosity and diffusivity in many solvents which are not ideal chromatographically but essential to dissolve the polymers). Columns are not as wide as are common in gel filtration, but bore, length (and particle diameter) are larger than usual for HPLC. High temperatures pose a different set of problems and make sample injection valves essential. Large peak volumes allow larger volume samples to be used and demands on detectors are less stringent as regards volume and sensitivity.

So, at present, surprisingly little equipment can be used for both techniques. There are signs that this situation is changing for workers in GPC are starting to take advantage of small particles whose potential is demonstrated by successes in HPLC. This can only be a slow process: reduction in time implies reduction in volumes and thus hardware changes. Because it is vitally important to measure absolute retention volumes precisely, some quite difficult improvements to existing pumps and volume measurement hardware are necessary. Bly et al[185] have stressed how important it is to avoid flowrate changes and how they lead to errors in the calculated results.

Polystyrene is by far the most popular packing. Because of problems packing columns and the need to keep columns wet with solvent and free of air, many columns are bought pre-packed. Proprietary materials have achieved a degree of *de facto* standard status and this is reflected in their price. But polystyrene is not ideal because it is not rigid and can stand only a limited pressure; it swells and deswells with changes in temperature and solvent; and is degraded at high temperatures and by air.

It is clear that most workers in permeation chromatography recognise the advantages of rigid porous particles[173], which usually implies an oxide like silica[178,179] or alumina[180,216] or porous glass[183]. For some non-polar polymers, typically polyolefins[182], silica is widely used for determination of molecular weight distribution.

Calibration

summary
- **calibration with materials of known molecular size essential**
- **absolute molecular weight distribution not easy**
- **comparison of results from different labs needs care**

Calibration is also a problem, for even if we knew the absolute size of the pores and molecules, these are not necessarily the effective size and shape in the presence of solvent. The effective size is called the hydrodynamic volume. Thus in practice systems are calibrated with molecules, usually of polystyrene, with a narrow molecular weight range, and whose molecular weight (and less precisely size and shape) are known from other techniques. Ouano reviews critically[187] the problems and progress in calibration. The key problem is that most detectors, typically RI, do not record molecular weight directly, which is what is wanted by most users of GPC for polymer analysis, and the standards for calibration are only a partial solution. Detectors that record parameters closer to the molecular weight like viscometry[189] or light scattering[188,219] are suggested as a way of reducing the problems of converting a permeation chromatogram to a molecular weight distribution. It is also essential to remove the instrumental peak broadening by some mathematical deconvolution to get an absolute distribution. This implies a computer system and because it is also convenient to use a computer to control the whole instrument, the trend must be towards on-line computer controlled chromatographs.

These problems with precise calibration do not reduce the usefulness of permeation chromatography as a method for comparing chromatograms of different samples run on the same chromatograph, but they do indicate the dangers of comparing chromatograms from different laboratories uncritically.

Permeation chromatography of polypropylene

CHROMATOGRAM
Column: Spherosil (Pechiney St. Gobin) 4-1000 nm
~1 ml/min 1-methylnaphthalene at 165°C
Injection : ~2 ml of 0.25% w/w solution
Detector : differential refractometer

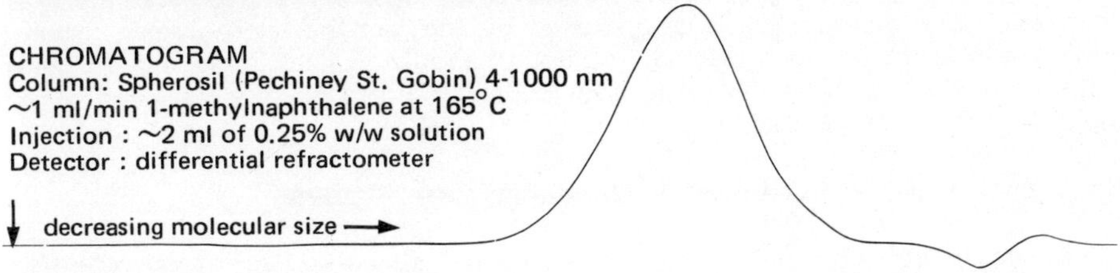

MOLECULAR WEIGHT DISTRIBUTION
(derived by computer calculation by comparison with whole polymer calibration chromatogram)

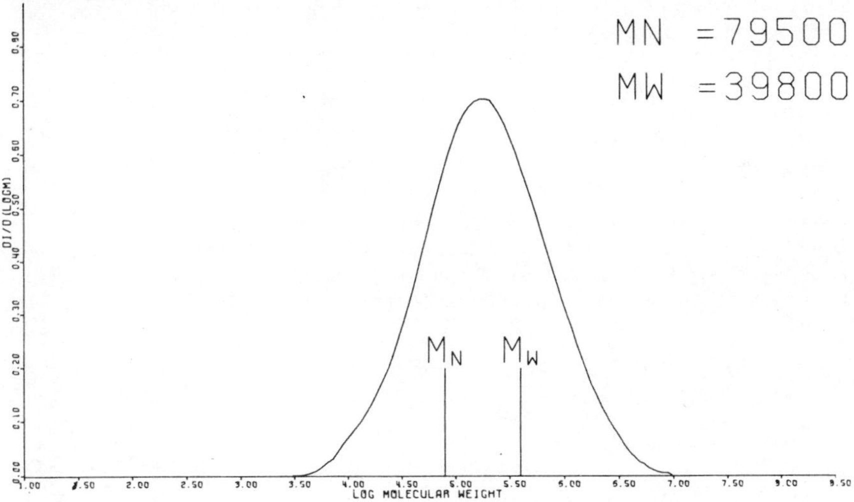

WESSLAU PLOT
(curvature shows deviation from gaussian size distribution; slope shows range of molecular size)

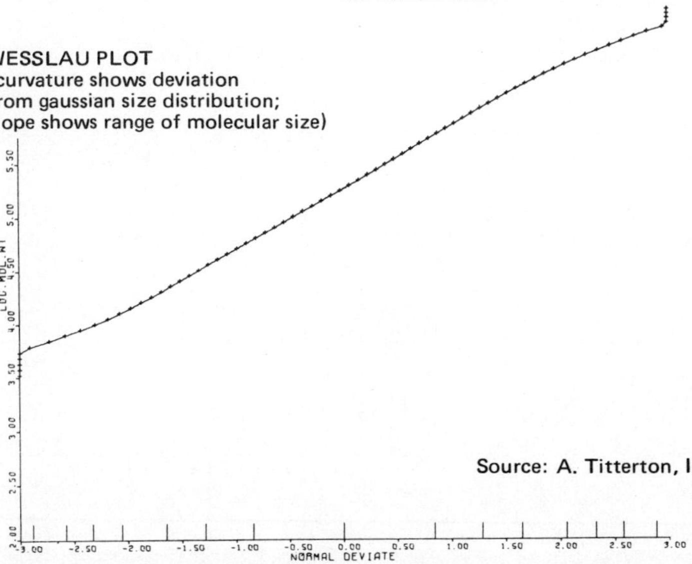

Source: A. Titterton, ICI Plastics Division.

Measurement of plate height (and hence resolution and other related parameters) in permeation chromatography is complicated by the fact the test polymer, usually polystyrene, is not a single molecular species but a range of sizes (molecular weights) and shapes. Bly[23] suggests that it is more realistic to normalize the plate number to the dispersity of the sample.

Effect of temperature in permeation chromatography

Unlike adsorption chromatography, the temperature has no effect on retention volumes[267] or therefore on selectivity. However the effect on plate number is much more significant than for small molecules which permeate freely. Higher temperatures lead to higher diffusivity. For example Unger and Kern[267] found the following plate numbers for a column 250 mm long packed with 10 μm porous silica and tetrahydrofuran as eluent

	24°C	100°C
excluded polystyrene	2600	2100
permeating polystyrene (molecular weight 98000)	150	240
ethylbenzene	6760	6900

So higher temperatures are worthwhile in PC to increase resolving power, in addition to the lower pressure drop (a factor of 2 for tetrahydrofuran by going from 24°C to 100°C). However the effective molecular size of the polymer may also change with temperature if the chains become more flexible.

6.5 Pellicular materials

summary • **obsolescent**
• **limited resolving power, speed and capacity**

A pellicular material is one with a solid, non-porous, non-absorbing core and a thin chromatographically-active layer on the surface.

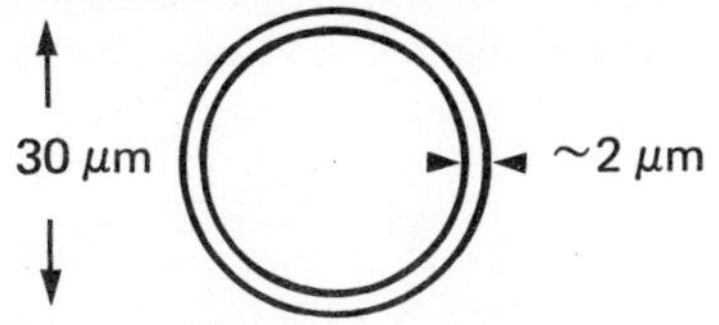

The rationale for designing such a column packing material was the realisation that if the *rate* of diffusion (mass transfer) was slow, in order to make sure it did not take too long, it was essential to reduce the distance molecules had to travel in the stationary zone down to a micrometre or two. Since it was considered difficult to make and pack particles as small as this, and it was known to be possible to dry-pack glass spheres of 30-50 μm, Horvarth, Preiss and Lipsky[42] and Halasz and Horvarth[43] decided to stick a thin layer of ion-exchanging resin onto impervious glass spheres. An important practical advantage of the solid core was that it would neither swell when wetted nor crush under pressure, two deficiencies of conventional organic resin materials.

Kirkland improved the surface with a commercially successful material called Zipax[339,340] which has thin layers of wide pore, low surface-area silica fused to the glass bead. This material has been the subject of detailed studies by Knox and co-workers: Kennedy and Knox[344] compared several materials of this type.

Separations achieved with pelliculars have provided a basis for development of many HPLC instruments and the surviving commercial types are quite robust and reliable. Because many separations are established on these materials, we can expect the materials to be in use for some years yet even if other materials can separate faster or better: too much separating power can be an embarrassment rather a blessing. Historically, pellicular materials lie at the end of an evolutionary branch and are destined to become obsolete.

Now that slurry packing methods are well developed (see section 4.2), the dry packing methods used for pellicular materials are inconveniently slow and need too much skill.

The flaw with the pellicular concept is that the solid glass particles occupy a large column volume and cause peak broadening, however well they are packed, but do not contribute at all to the separation. Peak broadening not only reduces resolving power, unless extremely long columns are used (see section 8.10), but dilutes the sample so that a high detector sensitivity is needed. And the weight of chromatographically active material is small, so only small weights of solutes can be injected, again increasing the need for high detector sensitivity. Pellicular materials are therefore unsuitable for preparative work, or for trace analysis.

Some simple calculations can show the considerable differences between pellicular materials and totally porous materials. If the diameter of the glass sphere is 30 μm and the silica layer 2 μm, then the fraction of the total sphere volume that is silica is only 0.3. Only about 1% of the total weight of a bonded-phase material is organic, so a resin, perhaps ion-exchanging, with 4 μm diameter particles may have similar properties, but a hundred times the capacity of a pellicular coating of the same material. This high capacity may allow a much shorter column to be used which will compensate for the lower permeability of small particles, and reduce the problems caused by the softness of the resin which leads to columns that block up at very high pressures.

Pellicular materials with bonded phases

Since the surface area of Zipax is only 1 m^2/g (compared to several hundred m^2/g for typical porous silica packings) it is almost inert, but an ideal support for other materials, bonded or not: in practice a polymerised octadecylsilane bonded-phase has proved most popular[345,346].

In order to get a practical capacity from an organic surface layer, more than a monolayer is needed. This factor, combined with the intrinsically low surface area, means that some of the problems of the surface showing through, found with materials bonded onto porous silicas, are avoided.

This factor leads to a robustness which will ensure their continued use for some time, even in preference to a higher performance, but shorter life, ODS on porous silica.

Other materials like Corasil (Waters Associates) have a thicker layer with a higher surface area (but at the price of less mechanical stability which makes it difficult to slurry pack without damage[236]) so they can be used as adsorbents. However the performance of porous microparticle silicas is so much better than that of this type of material that it is difficult to see any reason for continuing to use pellicular silicas.

The first phases to be coated onto the surface of silica were liquids like β, β-oxydiproprionitrile[341] which were held only by physisorption and not by a chemical bond. These are now little used because the stationary liquid dissolves in the mobile phase and must be continuously replenished from a pre-column and/or an eluent saturated with stationary phase. (Liquid–liquid systems are discussed in more detail in section 6.2.)

Chemically bonded phases are now well established commercial products. The stability of many early materials against removal of the 'bonded' layer by the eluent left much to be desired, but manufacturers have now improved their techniques by choosing only suitable bonding groups.

Alkylsilanes are by far the most popular group, judged by published methods, and octadecyl the commonest alkyl group. The degree of polymerisation varies from one manufacturer to another, for example being high on the low surface area Zipax used to make ODS-Permaphase, but much nearer a mono-layer on Vydac and

Bondapak, based on higher surface area supports. The degree of polymerisation is a source of batch to batch variation which manufacturers must strive to control. Resins with ion-exchanging groups like tetra-alkylammonium or sulphonic acid are also widely used and available from most manufacturers of pellicular materials. Organic groups with a terminal —CN are also bonded on by some manufacturers. Majors[214] reviewed the materials available in 1975.

summary • **robustness of thick polymerised layer is key factor**

advantages	**disadvantages**
can be dry-packed	usually dry-packed
	low capacity
low surface area — inert surface	high pressures usually needed
thick polymerised layer — robust	limited separating power
	limited separating speed

7 HARDWARE

Hardware too often receives obsessional interest. Blandishments of the marketing men must be prime suspects. Any obsession should be about the column, its engineering, contents, packing and performance. No separation occurs in the pump, detector, digital displays or Design Centre Award winning case, but to the detriment of chromatography these items are essential, and profitable.

Basically LC is quite simple and the hardware requirements can also be quite simple. Instruments designers must take some blame for the generally over-complex equipment now being sold for they have tried to meet every users demand for facilities and convenience. Users must also take some blame for buying. Complexity and price rise in proportion, but reliability is inversely related. Equipment has certainly received a disproportionate share of the development expenditure compared to that spent on column packing materials.

This chapter aims to help the user select the best equipment for his needs. These do vary widely and there is no 'best' type (or manufacturer) of LC instrumentation. The choice is certainly bewildering.

Complete Systems

A basic decision that has to be made is between a complete system from a single manufacturer or a 'homemade' chromatograph. Novices instinctively choose the security of the package; in practice the technique is still changing too rapidly for this security to last very long. The equipment for LC falls naturally into three modules that can be considered separately and chosen separately without problems of the interconnections being a real difficulty. If a complete system also meets the criteria of the individual modules it will obviously be a strong contender. Advantages of combining modules are in the ability to chose the best components and to create the most technically advanced systems; flexibility to accommodate changing technology; lowest cost, surprising but generally true; and most important, ease of diagnosis and maintenance.

General Scheme

Individual chromatographs, whether commercial systems or homemade combination of components, are all different but follow a general scheme. Solvent is prepared, perhaps a mixture formed, filtered and purified, and fed to the pump. The pump, with pressure measurement and protection, and pulse damping devices, drive solvent through the column via some injection system where sample is introduced. Separated components are detected and the results displayed, and the data processed. The liquid from the detector may also be collected so that the sample may be recovered, or pure solvent may be returned to the reservoir. All these components, and additional connections and facilities are described further in the following sections. In practice it is important to get the hardware to work well so that the chromatographer can concentrate on achieving separation.

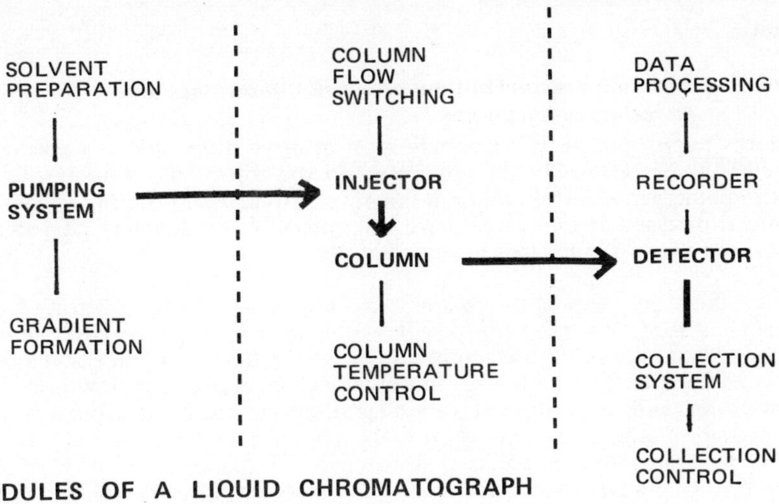

MODULES OF A LIQUID CHROMATOGRAPH

(key elements in **bold** type)

dotted barriers can be crossed by 1/16 in diameter tube

7.1 Minor Components

Solvent Reservoirs

summary
- **glass convenient for small volumes**
- **stainless steel safer for large volumes >2 l**
- **degassing, nitrogen purging, stirring useful but may be carried out before placing in reservoir**

Glass is convenient, transparent, corrosion-proof, and breaks. For small analytical systems it is ideal. Joints, if necessary, can be made with compression fittings with PTFE ferrules (but see section 10.5). 3/8 inch diameter glass tube is much stronger than 1/4 in. Even a short section of ptfe tube, 1/4 or 3/8 in diameter, between the reservoir and pump will buffer shocks. Larger glass reservoirs, above a few litres are less attractive because of the fire risk resulting from fracture. Trays to collect and contain the worst possible spillage will help to prevent or localise any fire. Stainless steel is preferable for preparative scale 20 l vessels. A sight tube, perhaps of thin-wall PTFE, is useful. It should not be possible to pressurize the vessel unless the arduous pressure vessel safety regulations (in UK at least) are met. Provision for vacuum and/or nitrogen purging can be met by a glass lid held in place only by gravity and vacuum. Nitrogen purging can prevent sample or column packing degradation, and also reduce dissolved gas for nitrogen is usually less soluble than oxygen. Magnetic stirring ensures homogenous solvent systems and helps degassing. It is also prudent to filter in all solvents. Although the solids content of solvents is low (or should be) quite large weights can be collected from the large volumes that may pass through a preparative scale machine.

Filtration

summary • **filtration essential but important not to restrict input to reciprocating pumps**

Dirt reduces pump reliability. Valves are most affected especially by fibres which do not clear easily. Material that filters out on the top of the column may also affect chromatography. Filtration should remove both particles and fibres. Sintered glass, porous ptfe and stainless steel have been used. Woven stainless steel mesh is also useful and shows a much lower pressure drop.

In practice there are some snags to the more obvious methods of filtration. The main problems arise from the very high flowrates for a short period when all types of reciprocating pumps refill. It is important to avoid a flow restriction or the pump will not refill completely, which leads to flowrate changes and to hydraulic hammer effects and larger pressure pulses. So very large filters are needed to avoid a pressure drop, especially if they become clogged with material they have filtered out. This then leads to a large hold-up volume, destroying an important advantage of small piston pumps. One alternative is to filter into the solvent reservoir which can take place as slowly as the average flowrate, but not restrict the input to the pump. A typical apparatus would consist of a three-neck flask accepting the pump inlet dip pipe, vacuum connection, and a glass sinter funnel.

summary • **pre-columns filter and purify and protect column**

Pump seals gradually wear and the resultant debris will collect on the column top, eventually increasing pressure drop. This too is best filtered before it reaches the column. At this point in the system a small filter will be satisfactory for the flow-rate is only the steady rate in use, and also some pressure drop can be tolerated on the high pressure side.

A pre-column is also effective at filtering and as a bonus highly polar contaminants in the solvent will be removed. Even highly purified solvents can turn a silica pre-column brown after long use. A pre-column will increase the life of columns used for repetitive analysis.

Degassing

summary
- **agitation controls rate of gas removal**
- **boiling, at ambient under vacuum, or by heating; vigorous shaking by hand, or ultrasonically; and pumping and recycling—all effective**
- **slow adsorption of gas so degassing outside chromatograph cheap and convenient**

Some very complex devices have been designed to degas solvents. Provided the detector cell is well designed to clear bubbles readily, only aqueous systems are likely to need degassing to avoid bubbles forming in the detector. An important exception is when two liquids are to be mixed to vary the eluent composition, when gas dissolved in one may be quite insoluble in the mixture and cause frequent bubble formation in the detector. Oxygen removal may be another reason for degassing, in which case vigorous sparging with nitrogen first is recommended.

Degassing is a kinetically controlled process. Heat or vacuum reduce the amount of dissolved gas in thermodynamic equilibrium but do not expel it in a reasonable time. Agitation is required to cause supersaturated solutions to quickly give up excess gas as bubbles. Boiling, either at ambient temperatures under vacuum, or hotter; vigorous shaking, perhaps by hand; ultrasonic agitation; pumping and recycling—all cause bubbles to form.

A simple but effective device consists of a 1 or 2 litre round-bottom flask with adaptor to connect to vacuum (a water ejector pump will suffice). A long rubber tube permits vigorous shaking while attached to the vacuum source. About a minute shaking is needed.

Less traumatic for the operator, since degassing is usually the first operation of the day, is an ultrasonic bath, as used (and most useful) for cleaning and dispersing slurries with which to pack columns. The flask, under vacuum, is partly immersed in the bath which is filled with water, detergent solution, or solvent used for cleaning: ultrasonic vibration is efficiently transmitted through the container to the solvent to be degassed. Quite soon, when no more bubbles appear, degassing is complete.

Degassed solutions absorb gas as slowly as they release it: they can therefore be stored or poured into reservoirs without adsorbing any significant gas. Only vigorous shaking with gas will cause solution. Even high nitrogen pressures cause only slow solution, as discussed in section 7.3 on direct gas-driven pumps. Avoiding stirring seems a plausible recipe for success (which leads to the speculation that adding baffles may be counter-productive).

Flowrate Measurement

summary

- **for application work measuring cylinder and stopwatch give 10% accuracy**
- **column performance appraisal needs 1% accuracy from burette and stopwatch**
- **permeation chromatography often uses more sophisticated systems**

Flowmeters are superficially attractive but do little more than indicate flow is switched on. The variable area and float type needs individual calibration for solvent or mixture in use, and have rather low intrinsic accuracy. Air bubbles and dirt cause wildly wrong results which may go unnoticed. For casual measurement applications work a stopwatch and measuring cylinder will suffice for 10% accuracy, but when a thorough column test is intended it is essential to aim for an accuracy of ±1%, for some column parameters are most sensitive to flowrate.

The simplest method is to use a burette or similar device with two marks a known volume apart and measure the time interval as the meniscus passes each mark. Obviously at least 60 s timing is desirable for this accuracy and the volume must also be known to 1%. Weighing the collected liquid over a similar time is also effective although the timing error may be greater than for watching a meniscus.

Bubble flowmeters can also be accurate and some automatic models are available. Siphoning devices are also automatic and convenient for larger columns and volumes and widely used in permeation chromatography where the absolute retention volume is important. In future computer–balance links will make feasible higher accuracy even for small volumes from higher speed permeation chromatography.

Pressure Measurement

summary
- **bourdon tube type useful but not ideal**
- **except when measuring direct nitrogen drive pressure**
- **electrical transducers useful for recording and automation**

The ideal LC transducer has a wide pressure range with 1% relative accuracy throughout; a few μl internal volume; clean sweeping flow channels; direct and electrical readout; good corrosion resistance; and reasonable cost. Alas few devices used in other pressure measuring applications come near to meeting all the needs and only a few specially designed and therefore expensive devices come near to meeting a few of these specifications.

Bourdon Gauges

These are widely and cheaply available but with a variety of annoying shortcomings:

Corrosion-resistant welded stainless steel tube types are rarely stocked.

Limited accuracy for column technology research as accurate gauges are not usually corrosion resistant.

Range over which reasonable accuracy can be read is limited so that several gauges are needed to cover the range 10 to 500 bar.

Air in the tube dissolves, so that pulse damping effectiveness reduces.

Internal volume changes with pressure reading.

Solvent trapped in the tube leaks out slowly when solvent is changed (though some versions have a valve to allow flushing).

High volume fittings unsuitable for LC are common.

Pulsing wear eventually wears out the mechanism.

The importance of these factors will depend on the application, and can be moderated by obtaining the best types, or isolating the gauge via a long length of small bore tubing. PTFE diaphragms widely employed in the process chemical industry are superficially attractive but too large and poorly swept to be practical for LC.

Pressure limit switches and linked potentiometers give cheap and convenient digital and analog readout.

Bourdon tubes have few disadvantages when they are used to measure the nitrogen pressure applied to direct-driven pumps. Precision types with bronze tubes can give very high accuracy over a good range and are ideal for column technology research.

Strain Gauge Transducers

Commercially available devices are not the ideal flow-through pattern. However the volume change with pressure is very small and they can be well isolated from the system liquid via a length of narrow bore tube. Most manufacturers adopt this approach. Even with this type it is difficult to cover the whole range of LC pressures accurately with one transducer, but the results are suitable for practical application work.

Column Temperature Control

summary

- **ambient temperature operation suitable for adsorption chromatography on silica and thin bonded layers**
- **thicker bonded layers, especially many pelliculars, and solid ion-exchange resins require higher temperatures**
- **permeation chromatography needs even higher temperatures to increase diffusion rates and solubility**
- **ovens are expensive, sometimes inconveniently sized, and their safety suspect**
- **circulating liquid jackets systems are flexible and cheap**
- **same liquid can thermostat pre-column and detector**

Column temperature control requirements change with column packing material development. The current trend is against the need for above ambient temperatures because the thin layers of chromatographically active surface allow rapid diffusion. Thus there is no incentive to reduce the solvent viscosity and hence increase its diffusivity. Reduction in the retention of the solute seems likely to negate any gains from increasing temperature. These factors were also true for the earliest separations using adsorption chromatography: thus all these were at ambient.

Partition chromatography using liquid-liquid systems, and later thick bonded layers, and the solid ion-exchange resins made higher temperatures essential to increase diffusion rates: thus all early HPLC instruments had column ovens or some other way of controlling the column temperature.

As surface layers exchange more efficiently, ovens have become more optional extras, further raising their cost. The small size of some columns has made choice of size impossibly difficult: ovens that accept 1 metre columns do not take 100 mm columns well. A more satisfactory solution is to water jacket each column individually from a constant temperature bath and a circulating pump. The initial cost is low (and many labs will already have one); temperature control good; heat transfer rate high; and one bath can service many columns; and it is far safer than air blown ovens where a small solvent leak can quickly give an explosive mixture even with nitrogen purging. Unlike GC there is no call for rapid temperature programming for the chromatographic effect of temperature is quite small.

Since it is difficult to jacket well at the column top near the injector, and yet here heat transfer should be good, a jacketed pre-column which also purifies and filters eluent is recommended. The constant temperature liquid can also be used to thermostat the detector. RI detectors make stringent demands on temperature stability; UV flowcells show less flow sensitivity if the cell and liquid are at the same temperature. Results are likely to be better than with two independent temperature control systems.

7.2 Pumps

Pressure

summary • **<100 bar needed to solve most problems**

The range of pressure used in LC varies from less than a metre liquid head used in classical column chromatography to the several thousand psi of some current work. There is a price for this pressure in the pump, plumbing and column needed. Several ranges, partly overlapping, can be distinguished.

type of equipment	approx. pressure limit	
	psi	bar
conventional lab glassware with conical joints and clips	15	1
larger glass columns with thick walls designed for LC	150	10
small bore glass tube with Cheminert or compression fittings, PTFE ferrules	500	30
stainless steel tubing and compression fittings (perhaps PTFE ferrules)	3000	200
as above but fittings are less reliable and fewer valves available	7500	500
special high pressure compression fittings and tube		2000

Prices tend to increase a fewfold from range to range. Since about 90% of separations can be achieved at reasonable speeds at <100 bar the need for higher pressures often implies inferior column technology. This combination can be an effective technological trade-off but needs to be specifically justified. Higher pressures also need to be used with an awareness of the problems that can arise from compressibility of the liquid and the temperature changes caused by the heat generated.

Flowrate

summary
- **analytical needs ~10 ml/min**
- **prep can use >200 ml/min**
- **not all pumps give this range**

Flowrate requirements vary very widely. 1 ml/min may be adequate for analytical work (though 10 ml/min would be a better target), but 200 ml/min is a reasonable requirement for preparative columns as small as 1 inch. The range of flowrate that a pump can give is the main feature distinguishing constant pressure from constant flow pumps.

Suitable materials of construction are vital. This requirement is difficult to specify for who knows what eluent will be selected next: it caused considerable problems to the pioneers of high performance liquid chromatography. Materials are discussed in detail in section 10.

Electric or pneumatic power is used to drive pumps. Constant flow usually results from an electric motor which often runs at a constant speed; constant pressure from a constant pneumatic pressure.

Pulses

summary
- **dual pistons pulse little but cost much more**
- **pulsing more or less undesirable**
- **dampers can reduce pulses**

Piston size determines the frequency and importance of pulses from refill strokes, and a great deal of effort has been devoted to removing these flowrate changes because of their detectors. Although detectors have been made much less sensitive to flow rate changes, all are still somewhat sensitive and it remains desirable to reduce pulsing.

There have also been fears, but little real evidence, that pulsing spoils the chromatography. It may not be coincidence that most of the very best results reported in the literature have been with pulse free pumps of various types.

This problem is most common with small piston pumps, and has become more serious as the cycle time, around a second or two, has come closer to the very short peak times of short and efficient microparticle-packed columns. Larger piston pumps usually accept refill as an activity to be carried out for preference between chromatograms.

Multihead systems where one piston continues pumping while another refills reduce pulsing to low levels. Some clever sinusoidal drive cam systems have been devised to keep flowrates constant. This adds rather a lot to cost and complexity. Alternatively a pulse damper can be fitted between pump and column. These are never completely effective and, more seriously, negate the main advantage of small piston pumps which is their small low hold-up and rapid solvent change.

The simplest pulse dampers consist of a tee piece with air trapped in a side-arm. A bourdon tube gauge acts like this (as well as showing the pressure). But slowly the air dissolves and effectiveness falls. A tap at the end of the tube can allow one to introduce new gas. To prevent gas solution diaphragms have also been added to this basic concept.

Valve Seats

PTFE is an attractive material for valve seats, for example in non-return or pressure relief valves used in LC pumps. The material is tolerant to particles which are bedded into the soft material and its flow property accommodates any roughness on the ball or poppet caused by corrosion. Correctly designed, the svstem pressure acting over a small area will cause the material to flow and seal. It is more difficult to obtain ideal sealing forces over a wide range of pumping pressures, and a stop to prevent excessive flow at high pressure and springs to seal at very low pumping pressures are needed in practice. Seals in which elastomers are replaced directly by PTFE are rarely satisfactory.

Most small piston pumps use ceramic balls and seats. These wear very little even operating once per second or so but are less tolerant of particles, especially fibres, so excellent filtration is essential.

Electric Pumps

summary
- **constant flow natural**
- **small piston synchronous motor cheap**
- **diaphragm avoids piston seal problem**
- **stepping motors can be controlled for flow or composition change**
- **large piston types pulse free**
- **limited max flowrate makes less suitable for prep**

These naturally give constant flow. Small piston types (1/8—1/4 in diameter) often have constant speed motors synchronized to the supply frequency. Flowrate is altered mechanically by adjusting the piston stroke length. This is simple and cheap but cannot be controlled easily by an external electrical signal.

Diaphragm pumps interpose PTFE or stainless steel and hydraulic fluid between the solvent and piston. This avoids piston seal problems, but diaphragm life is not unlimited. A limit can conveniently be applied to the hydraulic fluid pressure by allowing a relief valve to bleed fluid back to a reservoir.

The motor speed can also be altered through the drive pulse rate to a stepping motor. This can be used in a number of ways, for example to provide flowrate programming, or by using two or more pumps, to elute with a changing composition. The drive frequency can also be used to link flowrate and recorder x axis drive rather than recording against time, or used in some part of the data processing.

Small piston pumps can keep pumping as long as liquid is available without a long pause to refill.

Stepping motor drive is also applied to the large piston electric pumps. These do not produce pulses (or at least they are not detectable because of elasticity of various parts of the system) but eventually have to be refilled, quite a slow process taking a minute or so, long enough to interrupt chromatography. Again sophisticated control systems can control one or more pumps. Since the volume is ~500 ml, compressibility of the liquid and flexing of the seals cannot be ignored. This leads to errors with stopped flow injection as it takes time to build up pressure[31]. This can easily be overcome by raising the pressure rapidly to that determined for the flowrate set either before or immediately after restarting flow. Changing solvent is a slower and more difficult process than with small piston pumps. Designs must take the need to change solvent without unswept volumes to prevent removal of the old solvent.

A limitation of all electric pumps is on maximum flowrate. Electric motors capable of delivering hundreds of ml/min at ~100 bar are rather large, particularly if flame-proof for safety (most desirable). Pneumatic pumps relegate these motors to the compressor house and are often more efficient in any case. The range of flowrate is also limited, perhaps a tenfold adjustment in constant speed motor types and a few times higher for stepping motor systems. This limits the range of column diameters that can be used: the range from 2 to 20 mm bore (a hundredfold change in area) is barely covered satisfactorily.

Gas Driven Pumps

summary
- **naturally constant pressure**
- **direct—usually coil—reliable and cheap but inconvenient for solvent change**
- **pneumatic amplifier ideal for prep but expensive for analytical**

All gas driven pumps naturally give constant (or at least controlled) pressure. Drive can be direct, or amplified by a large air-driven piston pushing a small piston into liquid. Flowrate is limited by the pipework and column impedance.

Direct Gas-driven Pumps

A reservoir, often a coil, is filled with liquid and regulated nitrogen pressure used to drive the liquid out and down the column. Conventional cylinder-head pressure regulators are used to control the pressure and hence indirectly the flowrate. In practice an interlock linking the valve is essential to prevent incorrect and potentially hazardous valve position combinations, for example linking the high pressure gas directly to the reservoir while it is open to the atmosphere for filling. Pressures above about 2000 psi are unattractive because of safety fears and the limited life available from standard nitrogen bottles filled at 2500 psi.

The coil is now the traditional way to reduce the contact area of gas and liquid and thus limit the gas that dissolves to a short section of liquid near the interface. In practice gas solution is not as important as generally imagined. Objective measurement of dissolved gas has not yet been described: the results might reassure apprehensive potential users. Other configurations are also effective[58]. One was discovered by accident. To avoid the bulky coil a 1 m vertical tube 1 in diameter was used instead. Preconceptions about dissolved gas led to a design of a complicated glass float which was self emptying on depressurisation. This reduced the gas-liquid contact area to that of a coil type. After several months of successful use it was deemed highly effective—until the float was found in a drawer.

Direct gas drive is extremely reliable. This attribute is undervalued. It can give excellent pressure control and is absolutely pulse free. Low cost arises partly from low initial cost: more important is the minimal maintenance cost. A decade of operation with only minor adjustment can be safely predicted.

In practice this type is extremely convenient for routine analysis where the same, or similar, solvent systems are used often. But changing solvent is much less convenient and wastes some solvent, so separation development is slower.

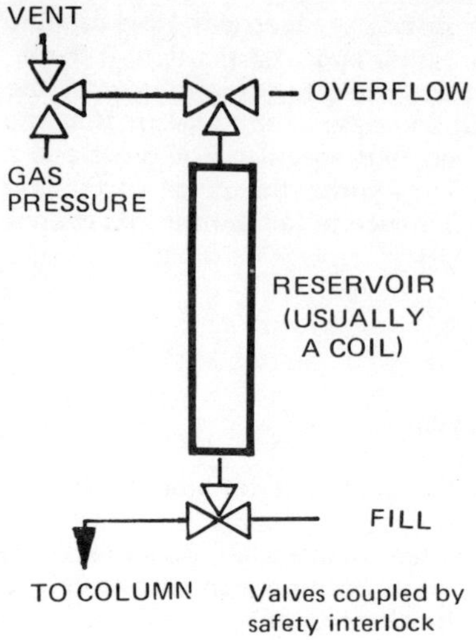

Direct Gas-driven Pump

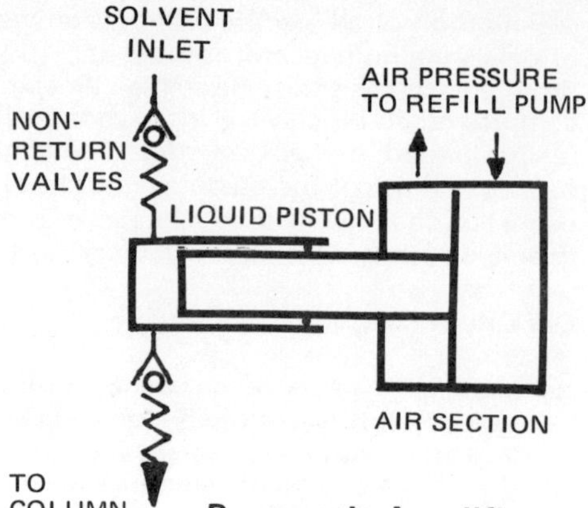

Pneumatic Amplifier

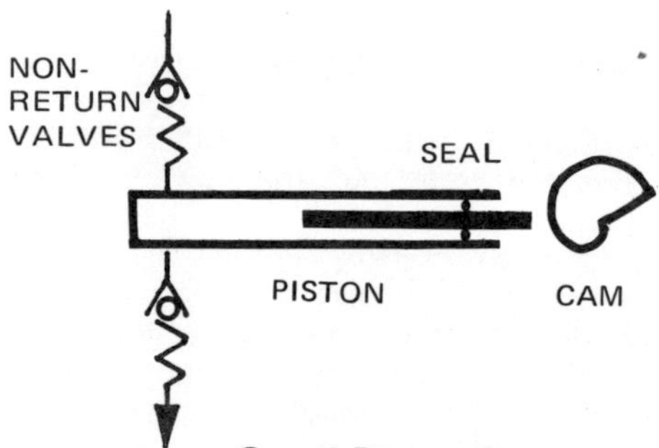

Small Piston Pump

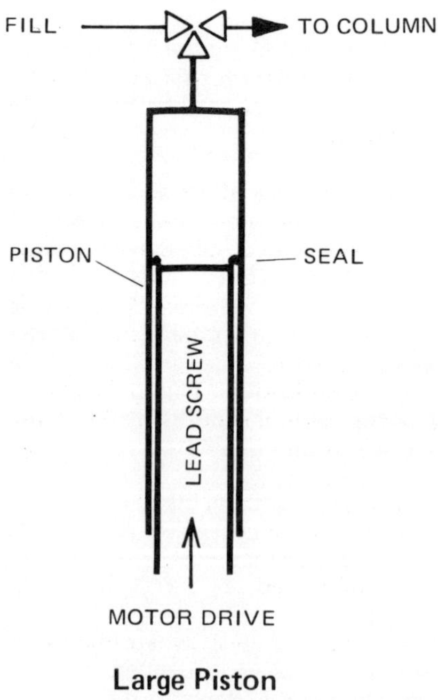

Large Piston Electric-driven

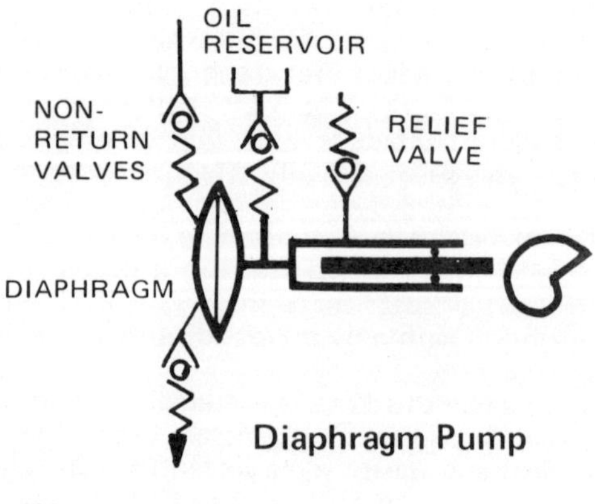

Diaphragm Pump

Pneumatic Amplifier Pumps

The pressure amplification ratio is determined by the ratio of area of air-driven piston to the liquid pumping piston. Ratios of 20 to 50 are common. Maximum speeds are determined by the maximum reciprocation rate, usually limited by the air supply pipework, and the volume per stroke which can vary from 1 ml to 100 ml. Pulsing obviously becomes a problem when high reciprocation rates are used. Miniature types with a few ml per stroke can still give 100 ml/min and large versions can give several liters/min which should be enough for chromatographers needs for the foreseeable future. At the other end of the scale, zero flow is achieved effortlessly and the flow may be shut off with a tap without further safety devices, an important practical advantage. Unlike electric piston pumps, especially large piston versions, the pressure and flow build up immediately the flow is turned on again.

Large diameter air-drive and pistons give better pressure control because of proportionately lower friction. Friction around the circumference is related linearly to diameter but force increases with the area and diameter squared. The air pressure required to overcome friction determines the range of useful pressures that can be generated. Less than 1 psi air may be required; thus air pressures up to 150 psi give a hundredfold pressure range. However larger diameters are more expensive; around 6 in diameter air section seems a reasonable compromise.

Constant Pressure versus Constant Flow

Arguments about respective merits continue.

Constant flow is certainly more convenient but in practice it is still necessary to measure flow for high accuracy. A number of factors, some physical like compressibility, and others mechanical like seal elasticity, lead to flows that are not precisely that set on the dial. Efforts to reduce these deviations by flow-feedback control have almost turned constant flow back to constant pressure types—at high cost.

Constant pressure too does not lead to constant flow because the column and pipework impedance changes. Pump seal and septum debris, and insoluble sample components are the main culprits for the impedance of well-packed columns is quite constant.

An entirely rational choice is impossible. Factors which are not central to this argument like cost, reliability, maximum pressure and flowrate, and gradient capability are probably decisive. The following recommendations are acknowledged as contentious:

- **coil pumps** for simplicity, low cost, high reliability, routine, high sensitivity and student use—and first-time users.
- **small piston** constant speed motors are cheap but less reliable and pulse—much better for frequent solvent system development.
- **variable-speed small piston** types allow low-pressure gradient formation (or high pressure gradient in pairs)—at much higher cost.

- **large piston stepping motor** systems are pulseless but only worth the high cost in pairs to form gradients.
- **large piston pneumatic** amplifier alone offers high flowrates for large preparative columns (but are also highly suitable for small analytical columns).

Pump Costs

Very approximate pump costs are:

coil	£ 500
small piston synchronous motor	£ 500
pneumatic	£1000
dual piston stepping motor	£2000
large piston stepping motor	£3000

Recycle

- useful only for mixtures of two very similar compounds (and no other impurities)

If it is intended to recycle eluted peaks back through the pump and down the column a second or more times then the hold-up and internal piston stroke volume must be much smaller than the volume of the peaks. Since the stroke volume of most commercial small-piston pumps (and most sensible designs) is ≈100 μl the volume of the peak must be much larger than those now emerging from efficient microparticle-packed columns. However recycle has been used to good effect in permeation chromatography where much larger and longer columns giving peaks several ml wide are common Martin, Verillan, Eon and Guiochon[388].

Safety of Pumping Systems

summary

- **different methods essential for constant pressure and constant flow types**
- **constant flow needs relief valve or electrical motor cutoff**
- **except diaphragm pumps with hydraulic pressure relief**
- **constant pressure best controlled via gas pressure limitation and max flowrate limited by choice of pipework bore**

Constant Flow Pumps

Blockages will lead to a pressure rise which must be checked or relieved, but leaks will not cause a flowrate increase. One approach is to measure pressure and set a trip level above which the motor drive is switched off. Another method is to use a valve that opens when the pressure is too high, or a bursting disc. All of these systems will introduce undesirable hold-up or even unswept volumes as do bourdon tube pressure gauges. Many types of relief valve used on process plant are far too large and often in unsuitable materials for LC. As a safety system bursting discs are most reliable but inconvenient. Since the consequences of a blockage may lead to damage but are unlikely to be dangerous, all these methods can be considered satisfactory. Indeed on pumps where seals are known to leak before damage is done no special precautions need be taken for reasons of safety, a chromatographically attractive solution.

Constant Pressure Pumps

It is poor practice to install overpressure trip devices on the liquid circuits of these types because pneumatic control is much more reliable and therefore safer. Conventional air pressure regulation and/or pressure relief equipment can be used to limit the maximum air or nitrogen pressure. For example the maximum air pressure to a pneumatic amplifier pump can be limited by an additional pre-set pressure regulator. The maximum liquid pressure cannot exceed this pressure multiplied by the amplification factor.

Direct gas-driven pumps should be designed with a safe working pressure above that which could arise if the regulator failed. In practice this means the cylinder pressure, often 2500 psi. Stainless steel tube, compression fittings and valves used in commercial designs meet these safety requirements easily and additional pressure relief systems are unnecessary and undesirable.

Flowrate limitation is, however, essential for safe operation of constant pressure pumps. This is simply and reliably met by restricting both gas and liquid flows by suitable choice of pipework bore. Direct high-pressure nitrogen should be limited by ~1 mm bore tubing (1/8 in diameter) and liquid flow on analytical instruments flow by 1/16 in diameter, 0.79 mm bore. These precautions are very easy. That they are obvious sizes to choose, and may well come with the instrument should not obscure their important contribution to safety.

Indirect flowrate limitation by air consumption control of pneumatic pumps is less straightforward without degrading pump performance in normal operation. Restricted feed to a reservoir large enough to supply air for the fast refill stroke is the only feasible approach. If the pump tries to run too fast this reservoir pressure will fall, but only quite slowly. In practice liquid flowrate reduction by suitable choice of output tube bore is simpler and sufficient.

7.3 Elution with Changing Solvent Systems

In function, eluent composition changes parallel temperature changes in GC. Components with widely different retention can be separated and yet eluted in a reasonable time. This is traditionally described as gradient elution, which implies a continuous change in composition following some complex function. A simpler system is stepwise elution which is worth considering further because it is cheap and easy whereas gradient formation is expensive and/or difficult.

Stepwise Elution

summary
- **simpler than full gradient but less powerful and not always applicable**
- **rapid composition change likely to provoke detector disturbance**

This simple solution is widely used in classical column chromatography but has been spurned by modern liquid chromatographers preoccupied with more sophisticated elution profiles. It is most useful when analysing components of widely differing polarities. At some point in the analysis, preferably when no peaks are emerging, the eluent polarity or strength is stepped up to elute quickly the remaining compounds. Compared to gradient elution the apparatus is simple. A single three-way valve can select a second reservoir to input the second liquid to a small piston pump; or two coil pumps can be selected in turn by a similar valve on the high pressure side. The system can be automated by activating the valve a set time after injection.

Unfortunately the rapid change in eluent nearly always produces a disturbance in the detector; this may be a refractive index effect in a UV detector even if the absorbances of the two liquids are well matched. A suitable time window in which to change eluent without confusing the chromatogram is essential and this means that stepwise elution is not always feasible in practice.

An important advantage for trace analysis, quantitation, and preparative work can be gained from the concentration of the highly retained compounds perhaps from dilute solution onto the top of the column.

Use of Stepwise Elution in formulated plant protection product Milcap®

Eluent: 1 hexane:dichloromethane 55:45
2 dichloromethane:MeOH:acetic acid 500:8.5:4.5

Detector: UV variable-wavelength (CE212)

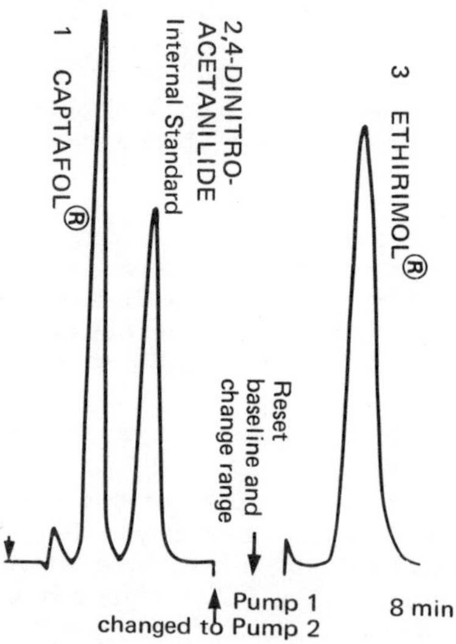

Source: R J Parker, ICI (Plant Protection Division)

Gradient Elution

summary
- **gradient powerful aid for separating components of widely differing retention**
- **gradient systems also convenient for separation development**
- **mixture can be formed at high or low pressure**
- **much more expensive than isocratic**

Gradient can be formed by mixing liquids at atmospheric pressure and then pumping this mixture with a single small-piston pump; or by mixing liquid from two pumps at high pressure.

Forming Eluent Mix at Low Pressure

A pump with a small stroke volume is obviously needed to preserve the composition profile. Stepwise selection from two reservoirs is the simplest example of this class and has been discussed above. More complicated (and some very complicated) systems have been described which feed one (or more) reservoirs to another from which the pump draws its feed liquid.

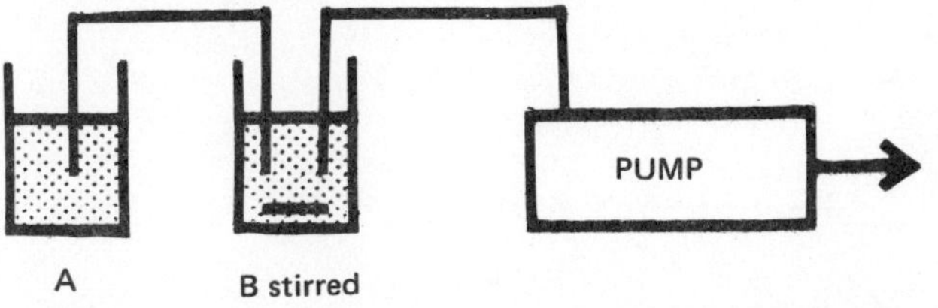

Alternatively a mixture can be formed by drawing liquid from two reservoirs which feed to a mixing chamber for a varying proportion of a short time cycle. (This time proportion control system is also used at high pressure and will be discussed in more detail below.) Actuation of these proportioning valves is usually electrical for they must operate very fast, especially for high speed chromatography. Control can come from mechanical or electronic timers, or follow the profile of a cutout former with a photoelectric sensor.

Good mixing is important, but since the pump also acts as a mixer it is less critical than in high-pressure time proportion systems. Detailed theory of this time proportioning system has been worked out by Keck[32]. The mixer volume, flowrate and cycle time relationship is important. A commercial version produced by LKB Instruments[33] has been widely applied to ion-exchange chromatography.

Gradients formed from three or more liquids, perhaps changing pH and ionic strength, are also conveniently formed at low pressure and fed to a single pump.

A number of other ingenious devices have been described. One by Perrett[374] is very simple and convenient for aqueous system, but poses some problems if the materials of construction must withstand organic solvents.

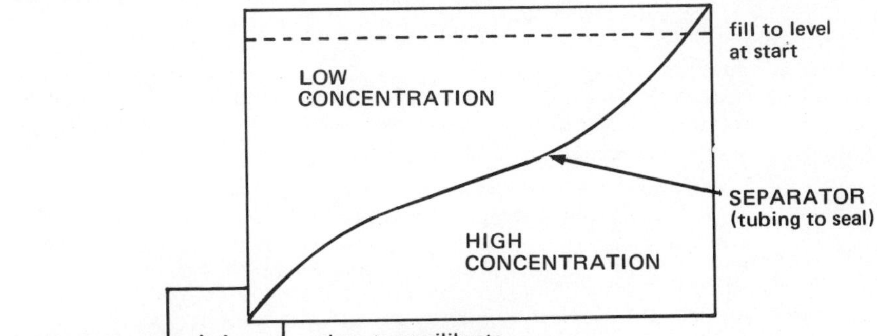

Gradient Formation at High Pressure

summary
- **two pump systems most common—convenient but expensive**
- **time proportioning feasible but valves critical**
- **mixing becomes more important for short efficient columns**

Two-pump Gradient Formation

The simplicity of taking two stepping-motor pumps and adding a single electronic controller has made this configuration most popular with users. It modularity is understandably popular with manufacturers. The pumps may be large or small piston types. The gradient is formed by altering the flowrate ratio to change composition but keep total flow constant. (Flow and composition programming is also conceivable.) Perhaps even more important in practice is the convenience of selecting an eluent composition at the turn of a dial, avoiding the chore of mixing solvents by hand. For separation development the cost of this type of system is easily justified but for routine analysis a single pump chromatograph will probably give better results at a fraction the cost.

Time Proportion Systems

Two constant pressure pumps can be used to create composition gradients by varying the proportion of the time, usually in a fixed time cycle, that two valves allow liquid to flow from each pump. Dupont Instruments use a variation on this scheme whereby only one pump is used with a coil holding liquid B but driven by liquid A from the pump. This avoids two relatively expensive pneumatic pumps but one must know when the holding coil is nearly empty: the equipment displays the volume used.

A more straightforward approach is to use two pumps (and there is no reason why more than two pumps cannot be used in an analogous way). Coil pumps are an obvious choice for they are quite cheap and the pressure will be exactly the same if a common nitrogen pressure source is used.

The key problem with time-proportioning is the valves. Excellent precision and good accuracy of valve timing coupled with very fast operation (milliseconds) are now needed to produce high speed gradients suitable for high speed columns resulting from the use of very small packing materials. Speed requirements dictate electrical actuation but this requires magnetic components whose corrosion resistance is less than ideal. The design dilemmas of time proportioning for high speed liquid chromatography have been discussed in detail[39] : the principle is sound but the implementation problematic.

Mixing

In the past when most gradients were of aqueous solutions and columns not too efficient, mixing the two liquids was, and is, a comparatively small problem: a simple tee is enough. When the two liquids, now much more likely to differ widely in density and viscosity, are mixed at high pressure and pumped down a short, low volume and efficient column a homogeneous mixture and smooth gradient can no longer be taken for granted: the efficiency of the column in not spreading solute zones also implies that it is a poor mixer. MacDonald[34] and Bristow[39] give examples demonstrating the advantages of power-assisted mixing. A short inefficient pre-column packed with large glass beads has also been found sufficient for long pellicular-packed columns using methanol-water mixtures, but is less satisfactory for more efficient columns.

The mixing problem is worst with time-proportion systems where a slug of liquid A is followed by a slug of B. It is fairly obvious that a stirred volume a little larger than the volume of one cycle, that is one slug each of A and B, is needed; Keck[32] presents the theory in more detail.

Composition Profiles

Most chromatographic problems require an exponential form; theory predicts an equal effect from each doubling of the concentration of liquid B in A. Two pump systems do not naturally produce this form for the speed control precision of each pump is more or less fixed at say 0.5%. Thus the precision of B at low B concentration is much lower than at high B concentration. For example it would be ±50% at 1% B in A, but ±1% at 50% B in A. This is unfortunate but to some degree inevitable. The problem is especially severe for time proportioning systems where the valve operating delay, a quite precise but comparatively long time, intrudes on the proportioning accuracy at very high and very low concentrations of A and B.

In practice many of these problems are minimised by choosing a mixture for the B pump which is the highest concentration of B in A that will be needed. Thus the B pump might contain a solution of 10% B in A giving a gradient from 0–10% B with much better accuracy than would result from using neat B. The mixing problem may also be reduced. It may also be more accurate to use three liquids; A, 10% B in A, and neat B, to cover the full range; this is catered for by the LKB Ultrograd design[33].

The implications of hold-up volumes in the pipework (and mixer if any) on the resultant composition profile should not be neglected. These may limit the maximum effective rate of eluent composition change.

In developing separations on a particular instrument, hold-up volumes and other profile non-linearities may be unimportant; but it is not safe to assume that a nominally identical profile on another model will produce an identical separation. This has implications for those specification analyses to be used internationally, perhaps to meet the requirements of regulatory authorities in the food or drug area. This is cause to eschew gradient elution unless essential: the best available column technology should be tried before resorting to the complication and cost of gradient elution.

7.4 Injection Systems

summary • **syringes troublesome but still have some advantages**
• **injection valves precise**
but difficult to achieve highest efficiency with microparticles

The theoretical aspects of injection volume and injection slug distribution are discussed elsewhere (sections 3 and 8.7). In practice injection is a difficult area. Many differences in column efficiency that have been ascribed to the column and its packing are in fact due to injection differences. The same column can give between 1000 and 10 000 plates depending on how the sample volume is introduced. This problem plagues interlaboratory comparisons, especially because, as will be seen later, the most efficient methods are also the most difficult.

The importance of good injection methods is well known for GC, especially for getting good peak shapes free from fronting or tailing[201]. (Theoretical formulae defining resolution rarely take account of the non-gaussian features of many practical peaks: a small front or tail can easily half the real resolution. Calculation of plate number from the width at half-height is convenient, but measurement of the width at, say, 1/50 of height might be more realistic.) Injection deficiencies show up most clearly for peaks that are only a little retained and may be less important in practice if all peaks of interest are well retained. However an improved injection method can permit a shorter retention time for the same resolution and thus faster analyses.

In LC the injection problem is easier than in GC because the sample does not have to be vapourised—but much more difficult because the volumes and distances are so much smaller. A further complication is the wish to avoid wall effects and therefore confine the slug to the centre part of the then 'infinite diameter' column. In GC the sample diffuses rapidly across the whole column area.

Injection systems are based either on syringes or valves. A number of hybrids have blurred the distinction but the clear difference that remains is that the sample flows directly from the needle tip into the eluent flowstream (whether on or off column).

Syringe systems can:

- pick up and inject very small volumes without any waste
- deposit the sample very precisely from the end of the needle
- quickly alter the volume injected from 0.5 to 1000 μl

but are:

- not very robust (and very fragile in some hands)
- difficult to use against pressure >100 bar
- wear out, leak, get blocked, and seize up
- an appreciable expense in running a chromatograph
- need some septum or other seal
- difficult to clean completely.

Syringes are therefore good for development but not so good for routine analysis.

Septa

summary
- **all materials are troublesome**
- **pays to select the best material for a given solvent system**

Many septum materials have been tried: all are bad, and some are even worse. Some materials are particularly resistant to some classes of solvents. Natural rubber (red) is good with aqueous systems, perhaps with polar additives like alcohols and acetonitrile, but quite useless with chlorinated solvents. It is quite tough and seals well. Synthetic rubbers can resist chlorinated solvents tolerably but some will not withstand ethers, ketones or alcohols, and they can be cut easily by the needle so more septum debris tends to block both needle and column. Fluorocarbon compositions have good resistance to many solvents but are hard and do not seal easily. PTFE coated septa have perfect solvent resistance, but only until the protective layer is penetrated by the needle. Chacun à son goût.

Syringes have improved since the early days of LC when only GC syringes were available: needle guides reduce plunger bending; stops prevent plungers being projected across the lab by the liquid pressure; closer tolerances reduce leak rates against high pressure; PTFE-tipped plungers stop leaks and prevent plungers seizing as solvent evaporate from sample solutions.

Septumless syringe injection systems

Many improvements to syringe systems have concentrated on replacing the septum seal. An early solution was to accept stopped flow and use some type of valve to allow the needle to reach the top of the column[202]. This is still a practical solution and widely used in many variants. Other systems allow injection with flow on. An example is described by Spaans et al[203] who compress a Kel-F ring to seal around the syringe needle. There is some evidence that the very best results are obtained with flow on.

Personal preferences for or against the many commercial implementations vary widely. Systems that do not allow the needle to reach the bed (off column) cannot exploit the infinite diameter effect. They also tend to dilute the sample with eluent before it reaches the column itself, like many injection valves.

All improvements on the septum system are expensive and complex.

On-column injection

summary
- **on-column more efficient but more difficult**

On-column injection allows one to exploit the substantial gains in separation efficiency (at least halving the analysis time and pressure needed) from the infinite diameter effect. Current implementations require more skill to achieve this and often have a lower reliability. It is for robustness (and to avoid guarantee problems if the user interferes with the column top) that most suppliers of packed columns press a sinter into the ends of the column. This is achieved at the expense of the users' resolution. And if the sinter does block it can only be removed by drilling it out, absolving the manufacturers of further responsibility.

A compromise solution is to use a thin disk of stainless steel woven mesh (section 10.7) as a divider between the column packing and a layer of glass beads, glass wool or PTFE sinter or wool. All septum debris and solids from injected sample is collected in the glass bead layer which can be discarded as necessary and replaced. The thin mesh disk can also be replaced easily. A plug of PTFE can be used to prevent the beads falling out if the column is laid horizontal. The syringe needle is pushed in until it just touches the mesh. A guide to ensure the needle tip enters the centre of the column is useful.

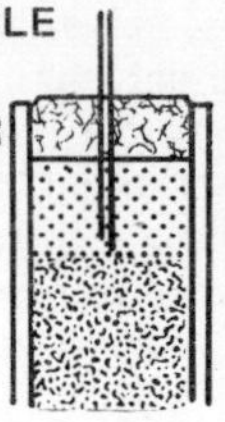

The mesh is also placed above the column tube in some designs[204].

Injection valves

summary
- **ideal for routine quantitative analysis**
- **improvements needed to realize full potential of microparticle packings**
- **most injection valves now dilute the injected sample too much before it reaches the column**

Valves designed for GC have not proved very suitable for LC because they are unable to seal at pressures needed for LC (though an ICI design of GC valve has proved reliable at 30 bar, a useful pressure for short columns packed with spherical microparticles[205]). Many commercial designs have become available in the past few years and their performance can now be described as satisfactory. Further refinements are urgently needed to realise the full potential of microparticle packing materials now available. In part this is a problem in how to connect the valve to the column so as to utilise the infinite diameter effect. Dilution of the injected sample before it actually reaches the column is also a problem with many valves, especially with small volume injections.

Important characteristics of injection valves are:

- precision
- can be automated (it is surprising that any valve cannot be remotely actuated)
- expensive to buy, but cheap to run
- can be used with flow on at $>$100 bar

but valves have drawbacks:

- difficult to change volume quickly
- dilutes sample to larger volume before reaching column
- most current systems reduce resolving power of combined injector and column system

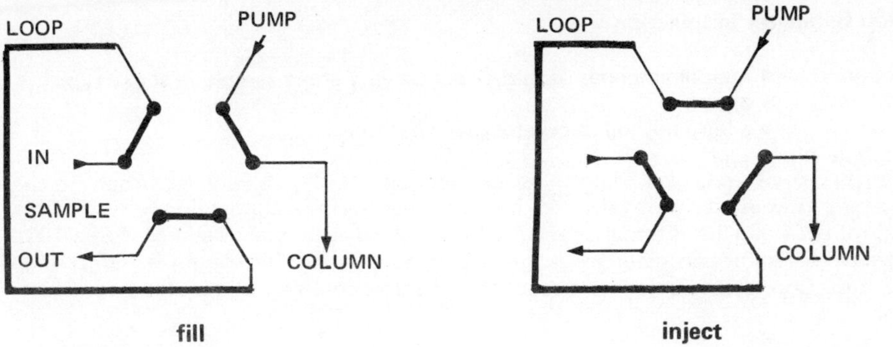

Sample injection valve

Some valves can be filled from a microlitre syringe, if necessary filling only a small fraction of the loop. This overcomes one of the major problems of valves, the sample size needed. There are several ways of implementing this but the basic concept is a major advance.

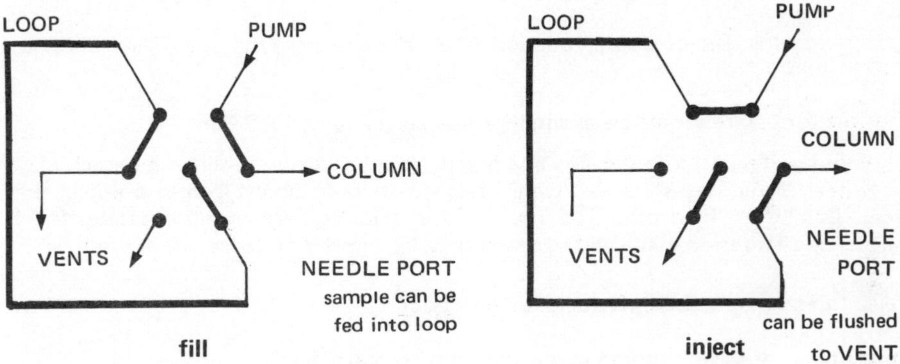

Syringe-loading sample injection valve
(Rheodyne model 7020)

Despite the importance of the actual volume into which the nominal injection volume of, say, 10 μl is diluted, data on the subject is almost non-existent. This parameter should form part of the specification of every valve. A recent scurrilous advertisement[206] provides a little data on this dilution for two commercial types. It proves deficiencies in either the injector, or detector, or both. But clearly the actual volume in which the injected volume reaches the column is much more than the nominal injected volume.

Loop Geometry in Injection Valves

summary • **sample chamber should either be very short sample in slider type**
or
• **a loop as long as practicable ≈0.3 mm**

The theory of zone spreading in tubes (section 12.4) is barely applicable to the complex flow patterns of valves[227], but it is clear that the connecting tubes, fittings and most of all, the internal channels, of many valves are far too large. For a 10 μl injector a very rough estimate suggests that no tube or slot should be larger than 0.3 mm. Connecting unions are suspect and some could be eliminated with advantage.

When designing an injection valve it is not obvious whether the sample chamber should be a very long and narrow tube, or a shorter one with much larger diameter. Tentatively theory suggests that either a sample chamber should be roughly spherical, when it will behave like a stirred vessel and deliver a fairly compact, but exponentially tailing slug, or it should be as long as practicable without causing too much pressure drop. The first type of geometry is typical of the sample in slider configuration, and the latter of loop injectors.

The latter conclusion is reached by considering the diameter of tube which gives the least broadening for a given volume.

Since, from the Taylor-Golay equation $\sigma^2 \propto d^4 z$ and $z \propto 1/d^2$, for a given volume, $\sigma^2 \propto d^2$ and $\sigma \propto d$.

Thus the loop bore should be as small as feasible, usually ≈0.3 mm.

For volumes <5 μl, loops become less practical and a sample-in-slider approach may be better. Many valves for GC were designed to hold about 1 μl in a slot in the slider, usually of filled ptfe. The Taylor-Golay equation is a less appropriate model for this configuration and a stirred vessel may be a better basis for calculation.

Valve Injection and the 'infinite diameter effect'

summary • **valve—column connection has critical effect on efficiency and peak shape**

It is not easy to obtain reduced plate heights of 2–3 with valve injection, even on columns that have been shown to give this efficiency with syringe injection. One reason may be the sample dilution caused by the valve, but another is the difficulty of connecting the valve to the top of the column so as to operate in the infinite diameter mode[106]. The most obvious method is to use low dead-volume fittings similar to those used for column end fittings. (Low dead volume is much more important at the injection end than at the detector connection because the 'peak' is perhaps a tenth the size it will be after it has passed down the column.) Some more complicated arrangements aimed at maintaining coaxial flow with a curtain of solvent down the walls have given higher efficiency and, more important, better peak shape[205] but further investigation is still needed to find the most practical method of implementing this system.

Assessing injection systems

Factors worth considering are:

- can it be used with flow on, off, or both
- on or off column—that is, can it exploit the infinite diameter effect
- memory or carry-over
- range of volumes that can be injected, and whether new parts must be installed
- dilution of the sample slug
- cost, reliability and availability of spares
- can it be automated (almost invariably pneumatically actuated)

Memory effects can be tested by alternately injecting sample and a blank, say, the solvent used to dissolve the sample. It is desirable to use two syringes for this test because it is difficult to be sure that a syringe is completely cleaned: filling a syringe with red ink and trying to clean it will demonstrate the problem. Needle guides, septa and unswept volumes are all sources of memory effect.

To assess dilution of sample requires a very small volume detector: detectors are normally used to detect injections that have been diluted to a larger volume by the column. To assess a nominal 10 μl injection volume, a detector volume of 2 μl is desirable (one of 8 μl is clearly not very suitable).[226]

7.5 Columns

summary • **stainless steel columns (with compression fittings) most popular**
• **glass useful at lower pressures**
• **connections for injection and detection are important**

Most workers are now using stainless steel tube for columns with proprietary end fittings, many based on standard compression fittings (see section 10.6) but modified to retain the packing and avoid dead volumes which re-mix the separated peaks.

The need to pack slurries at quite high pressures has made glass less popular despite its naturally smooth finish and excellent resistance to corrosion.

Surface Finish of Column Bores

summary • **smooth bore improves performance of long thin columns**
• **surface finish less important for infinite diameter columns**

Kirkland[114] and Majors[115] have both shown the importance of a smooth bore when using non-infinite diameter columns, for example using ~40 μm pellicular packings. Precision bore stainless steel, glass and ptfe coated tube all worked well. The evidence points to a good surface finish rather than geometrical precision of the bore. The internal finish of tubing to ASTM A269 varies widely from mill to mill, but precision bore tubing has a much smoother bore. Polishing the bore with abrasives is feasible but tedious. Peterson[102] has recently described electro-polishing[112] of the bore of stainless columns: the process can be applied to long (1 m) and narrow bore (2–3 mm) columns. Improvements produced by drilling, which produces grooves around the circumference rather than along the length are also reported[113]. Wall effects persist with small particles according to Laub[117].

With larger columns ~5 mm and small particles $\leqslant$10 μm, and central on-column injection giving operation in the infinite diameter mode, wall effects are less apparent and there is little evidence to suggest that polished tubes are worthwhile in practice.

For all this data we are still ignorant of the effect of the type of surface roughness and its interaction with particle size and shape, and the interaction of all these parameters with packing method. It is plausible to suppose that irregular particles, with a fair proportion of very small particles, packed from a dilute slurry might fit into the wall roughness, but we still do not know how much dispersion takes place near the wall. The methodology used by the above workers, and very many others whose results are never recorded, is inadequate for gaining a detailed understanding of the interaction of the many variables: but essential to establish whether a particular combination of finish, packing type and size, column geometry and injection distribution system will give useful efficiency and solve a real problem.

7.6 Detectors

summary
- **fast monitoring essential for small particles**
- **'universal detector' unlikely**
- **photometric detector widely applicable, and popular**
- **RI and other bulk property detectors barely sensitive enough**
- **other detectors remain in development phase**
- **novel detectors still needed**

An essential part of a modern liquid chromatograph is a continuously monitoring detector. A change from manual point-by-point monitoring is dictated by the higher data rate and small volume, and amount, of each sample. Microparticles operated at their minimum speed require data collection at rates equivalent to several points per second: resolution is lost at lower speeds. Peak volumes are often less than 100 μl indicating that the 'industry standard' of 8 μl is none too small. Larger and/or slower detectors throw away some of the resolution achieved by chromatography.

Sensitivity in an LC detector requires extremely high selectivity to detect minute concentrations of sample and yet be insensitive to a vast excess of eluent, perhaps of varying composition. These requirements rule out almost all physical properties as a basis for detection, and make the search for a Universal Detector quite hopeless. Measuring a bulk property such as refractive index, or dielectric constant, and infering solute concentration as a very small change (say 10^{-6}) is difficult and even the best devices are barely sensitive enough.

The nearest approach to a universal detector is made by the photometric detector which is significantly the type used by the pioneers of modern LC in the late sixties. The performance of this class is good enough to make it difficult to foresee more than a tenth of applications using other systems unless a dramatically novel alternative is developed. UV photometric detectors now occupy a position analogous to that of the flame ionisation detector (FID) in GC. Carrier gas is much more different from the sample in GC than the eluent from the solute in LC.

Sensitivity

summary
- **sample sensitivity may differ from absolute absorbance sensitivity**

Absolute sensitivity compares the signal output with noise. For photometric detectors it can be taken as twice the peak-to-peak noise, that is signal to noise ratio S/N = 2, giving an absolute sensitivity in absorbance units terms. This provides a figure of merit basis for comparing instruments, and their specifications.

Relative sample sensitivity is the minimum amount or concentration of sample which will produce a given signal/noise ratio, usually 2.

Examples where these two sensitivities are very different in practice are in photometric and refractive index detectors. If two detectors with equal absolute sensitivity of 0.001 A are compared for monitoring a compound with a hundredfold greater absorptivity at 220 nm than at 254 nm the relative sample sensitivity monitoring at 220 nm will also be a hundred greater. Similarly the sample sensitivity of refractive index detectors can be changed manyfold by a change in solvent—solute refractive index difference.

It may not be easy to answer the question—"is this detector sensitive *to detect my sample*?" For photometric detectors one can easily calculate the minimum concentration that can be detected for a given pathlength from a spectrum of a sample of known concentration: it is more difficult to know how much dilution will occur as the sample passes down the column. Dilution might vary from as little as tenfold for large injections on short microparticle-packed columns, to as much as several hundredfold on long columns packed with pellicular materials on which only small injections are possible.

Noise

summary • **only noise in frequency range of measurement significant**

Noise is random signals interfering with the measurement: the frequency of noise compared to the bandwidth or frequency range of the measurement determines how much noise interferes. Noise at radio-frequencies above 1 MHz has negligible effect on measurement of the slowly varying chromatographic signals; baseline drifts over periods of weeks which are much lower frequency noise, say 1 μHz (1 fortnight $\approx$1 Ms and 1 μHz $\approx$1 fortnight^{-1}), can also be neglected.

Noise in the frequency range of measurement is important. Conventionally, and usually rather vaguely, noise at frequencies at the top end of the measurement data bandwidth is called 'short-term noise'; and that at the bottom is called 'drift'. To be pedantic the noise frequency range should be stated. In chromatography, short-term noise usually means above 1 Hz (and up to recorder response speeds of about 20 Hz) and appears as 'grass' on the trace, whereas drift is below 10^{-3} Hz (1 mHz) with a period of hours rather than minutes; long-term noise is at intermediate frequencies. For high speed chromatograms complete in minutes, drift is hardly important, but short term noise limits sensitivity and measurement accuracy. Slower chromatography with overnight running can permit one to electronically filter out or average the short-term noise but drift affects the results markedly.

Noise amplitude can be stated in terms of actual peak-to-peak value or as percentage of fullscale on a particular range, or deplorably often, in undefined terms. The '±' representation to make noise look half as large is also popular with manufacturers specification writers. Thus noise specifications in commercial and academic literature need interpreting with care and the impact of noise varies with the speed of chromatography used.

Linearity

summary • **noise and stray light limit overall linear range**
• **error in zero transmission setting leads to non-linearity**

The size of the smallest quantity that can be detected is important but the size of the largest matters too. In practice there is usually an increasingly non-linear region before this largest detectable concentration level is reached. This leads to definition of a linear dynamic range. The tolerable degree of non-linearity should, of course, be defined. More often a vague 1% non-linearity is presumed and the resultant data have a rather approximate meaning. In any case non-linearity may be tolerable provided the response can be calibrated precisely enough. With photometric detectors the logarithmic relationship of amount of light transmitted and concentration (absorbance) leads to rapidly increasing noise or reduced precision together with non-linearity from stray light. Thus the linear range of a <1% stray light (2 A) monochromator is double that of a 10% stray light (1 A) 254 nm type if they have the same noise levels. (But a lower noise 254 nm type might easily still have a greater overall linear range.)

Like sensitivity there is a difference between absolute and sample concentration linear ranges. Changing wavelength to where a minor component absorbs strongly and the major component weakly can allow the ratio of very different concentrations to be measured precisely.

For highest accuracy the chromatographic conditions, wavelength for detection, sample size and flowcell pathlength should be chosen to place the expected range of concentrations within the range of best linearity.

Response Speed

summary
- **microparticles need detectors to measure 1 second peaks**
- **still faster detectors not now cost effective**

High speed chromatography needs quite fast detectors. On short columns, say 100 mm long packed with 5 μm particles, modest reduced velocities produce peaks with time widths of about a second for unretained compounds. This is as fast as convenient in practice for pen recorders: devices capable of higher speeds and good accuracy are much more expensive. Photometric detectors operate well at this speed but transport systems and radioactive detectors tend to be slower. Transport systems may be improved by more sophisticated designs but fundamental physics of decay controls radioactivity detectors even if scintillation efficiencies can be improved.

Concentration or Mass Sensitive detectors

summary
- **most LC detectors photometric, and concentration sensitive**
- **transport plus FID or MS mass sensitive**
- **important implications for quantitation**

Unlike the FID traditional in GC most LC detectors are concentration sensitive. Thus if flow stops the concentration stays constant whereas the response of a mass sensitive detector falls quickly to zero because no new solute mass appears. Systems that separate solute from solvent and transport it to a FID or mass spectrometer are mass sensitive. Photometric detectors are concentration sensitive. This difference has most important implications in quantitation which are discussed further in section 8.4.

Characteristics of detectors

- noise (and how its frequency spectrum compares with that of chromatographic information)
- speed of response
- precision (repeatability)
- linear range of response, and total response range
- absolute sensitivity, for example to absorbance or RI change
- specific sample sensitivity
- concentration or mass sensitive
- selectivity or universality
- destructive or not
- does it give identification information such as spectra
- flow, temperature and pressure sensitivity
- sensitivity to eluent composition, especially for gradient elution work
- limits on solvents that can be used, for example UV transparency requirement
- zone broadening (before during and after flowcell) and sampled volume
- reliability
- cost, and running expenses
- operability

7.6.1 Photometric detectors — mainly UV

Providing the sample to be detected absorbs somewhat above 200 nm it is usually possible, if sometimes troublesome, to find a chromatographically suitable eluent which has at least a thousandfold less absorptivity. (Unfortunately there are a few large classes of compounds which do not absorb enough even at these short wavelengths, so there will always be a group of chromatographers calling loudly for other types of detector. Hydrocarbons are the largest single class; luckily most can be analysed by GC.) This insensitivity to solvent is the key to the success of this class of detector. However it also limits the wavelengths over which photometry can be used. In the ultraviolet (UV) beyond 220 nm it becomes increasingly difficult to find suitable eluents which do not absorb. In the near infrared (IR) the absorptivity of sample falls so that solute detection sensitivity becomes impractically low. The chemically interesting IR beyond 2 μm wavelength is promising but solvent absorption is rather too high compared to solute absorption to promise a practical detector with wide applicability. Sources and detectors are not as stable or sensitive as in the UV and it is not easy to produce a useful device with a high sensitivity and wide range. Since IR spectra are quite structurally informative a device that scans to produce spectra is attractive, but without improvements in the spectrometer, and additional data processing, the prospects are not good. As with LC-MS the fact that collection and sample recovery can also give identification information reduces the incentive for development of a complete detector as such.

Sources

Sources either emit a continuum of wavelengths like the deuterium and tungsten lamps; or distinct emission lines like the most popular mercury lamps. Some form of wavelength selection is still required unless a truly monochromatic source like a laser is being used. (Laser sources would be ideal for LC, where a minute aperture is needed, if only they were available at suitable wavelengths. Cost, reliability, and output amplitude stability leave something to be desired at present but improvements can be anticipated and one can safely predict their eventual application.)

summary
- **continuum sources used with monochromators**
- **fully variable wavelength**

Monochromators, usually grating types, are customary for continuum sources, especially deuterium lamps; interference filters for the short wavelength UV are difficult to make and expensive.

Stray light

summary
- **light outside selected wavelength bandpass**

If the filter does not remove light at other wavelengths then a non-linear response may be obtained at high absorbance if the compound being detected does not absorb equally at these so-called 'stray light' wavelengths. This misleading term lumps together any light not at (or within the bandpass of) the wavelength selected. The term arose from stray reflections in monochromators which are one source of the net light detected when all should be absorbed by the sample. Stray light is easily measured approximately by placing a filter or solution which should absorb completely at the selected wavelength(s) but as little as possible at others. For 254 nm mercury line systems use a solution, or injection, of toluene that would give a reading of at least 10 absorbance units. It is convenient to inject 1 μl of a suitable solution that gives a peak height between 1 and 1.5 absorbance and then inject 10 μl. The peak absorbance reading should be above 2 for a good instrument, which corresponds to $<1\%$ stray light; some instruments can give a reading of 4; a reading of 1 absorbance unit indicates a poor 10% stray. It may be necessary to adjust the recorder span setting to display these high values. A low value may also result from limitations in the transmission to absorbance converter. A precise statement of stray light (unlike that which features in most instrument specifications) needs a statement of the method of test[77]. This is because ideal filters (whether they be solutions, solid or coated glass) are never available with bandpass identically opposite to that expected from the filtered source. Practical test filters such as holmium glass give a useful comparative figure of merit.

As a rule of thumb one should ensure that peak heights are 1/2 absorbance less than the maximum reading obtained in a stray light test. In case of doubt a linearity test is the final arbiter and proves whether a system is accurate enough in practice.

Detector Volume

summary
- **detector volume should be $<$1/5 to 1/20 peak volume**
- **increasing column diameter increases peak volume which reduces demands on detector**

The volume is only important in as much as it reflects the parameter of real interest: how well does the detector signal represent the concentration profile of the actual peak. The volume in the flowcell determines the minimum volume that can be sampled but lack of ideal flow through the cell adds markedly to the zone spreading caused by the cell. Most of the zone spreading is asymmetric and causes tailing.

As a rule of thumb the volume alone of a flowcell used in photometric detectors should be between 1/5 and 1/20 of the peak volume. This can be calculated from the chart record, knowing the chart speed and flowrate; the computer program in section 12.3 also calculates the peak volume for the sample peak and the eluent peaks (the smallest possible). Above 1/5 peak broadening is very marked, below 1/20 further improvements are quite small. Large diameter cells cause much more broadening and tailing because of the volume in which mixing can take place and the poor flow sweeping properties of this configuration. This is unfortunate because large diameter apertures allow more light through to the photodetector. Calculation of the flow in cells is not straightforward; nor has a definitive experimental study of flowcell zone spreading been published yet. One problem is that of producing a small and well-defined testing pulse of sample. Successful use of existing designs is a further disincentive. Another potential testing method is to pass a peak successively through two or more flowcells and then rearrange the order of the cells. Mathematical analysis is more complex. The search for higher speed and more efficient columns, and better peak shapes, and the need to test sample injection devices more precisely, may provide an incentive for this type of study.

The volume of the peak is just as relevant as the detector cell volume. The peak volume depends on column efficiency, length and diameter. Larger columns relax demand on detectors, but small particles packed efficiently produce smaller volume peaks. If improved columns reveal detector limitations (seen most sensitively in the reduction of peak heights of early peaks compared to later ones) increasing the column diameter may help. Sample volumes will have to be increased in proportion to column cross-sectional area. Length increase also decreases sensitivity so must be the second choice method of increasing peak volume.

Theory points to considerable advantages in maintaining linear laminar flow from the column end rather than feeding liquid to the detector via long narrow tubes, but it is difficult to see how to engineer the column end arrangement into the detector so that the optical path is directly below the mesh or sinter retaining the packing. Its special attraction lies partly in avoiding detection of most liquid subject to wall effects and concomitant reduced resolving power.

Flowcell Design—some practical details

summary
- **bubbles clear quickly with good design**
- **flow impedance downstream pressurises cell and reduces bubble formation**

Gas bubbles that stick in the flowcell are a cause of severe user frustration. Good design of the flow pathways has reduced the problems found with some early designs. It also helps to be able to keep the cell under slight pressure for this keeps dissolved gas in solution. Several methods can be used: the simplest is a length of very small bore tubing downstream from the cell, for example a few hundred mm of 1/16 in diameter PTFE tube with 200 μm bore will give a bar or so back pressure. at analytical flowrates.

It is also convenient to be able to reduce the pathlength to alter the linear range especially when using LC preparatively. So cell designs with different cells or different inserts can have practical value. Zone spreading may be expected to be less for straight through cells feasible for 1 mm pathlengths compared to 'Z' or 'double T' configurations. Longer cells, say 100 mm pathlength, are also attractive as a way to increase sensitivity but the very small hole required to keep the volume down means that much less light will be available. (A tenfold increase in length reduces the aperture area tenfold.) A smaller source, perhaps a laser or light emitting diode, might ease this difficulty. It is also impossible to drill this long thin hoie straight and an inserted tube will be necessary to construct a cell of this shape. High flow sensitivity from the multiple internal reflections might also be expected. On the other hand in the UV some losses due to penetration of the UV-opaque wall on total internal reflection (as utilised in attenuated total reflection (ATR) spectroscopy) will accumulate on multiple reflections to attenuate light not travelling parallel to the cell axis and this may reduce flow sensitivity.

Flow Sensitivity of Photometric Detectors

summary

- **temperature changes in cell alter internal reflection and cause flow sensitivity**
- **equilibration of cell and liquid helps**
- **flow sensitivity less important with pulseless pumps**

Unexpectedly, all photometric detectors turn out to be flow sensitive. This is not a small effect: the transmission can change by over 1% as the flow is switched on and off. In practice it has often limited the ultimate sensitivity of UV detection.

Unlike RI detectors, temperature and pressure can be quickly eliminated as primary causes, for absorptivity is only very slightly affected by these factors. Using a long and narrow flowcell produces the symptom. After some time an explanation of this puzzling phenomenon appeared[110]. In the cell many internal reflections take place because the optics do not produce a parallel beam. These internal reflections are very sensitive to changes in refractive index near the walls. Thus if the temperature changes as the liquid passes down the cell the pattern of reflection will change and so will the effective pathlength and reflection losses. More or less light will exit outside the collection solid angle of the photodetector.

An important reason for the diverging beam is that the source image is usually focussed on the cell to maximise light throughput, and the cell is used as a light pipe. With visible light the exit beam can be seen as a series of many overlapping circles, images of the cell aperture. Some recently designed instruments produce a more nearly parallel light beam and thus fewer reflections.

All designs are improved by ensuring that the cell and liquid are at the same temperature. Equilibration not thermostatting is required. Keeping hot lamps away from the cell is a simple step and gives monochromator types a natural advantage. Equilibration systems, usually a coil of narrow bore tube in good thermal contact with the cell, can help (but can cause extra zone broadening if clumsily designed).

Callmer and Nilsson[209] describe the improvement to the practical sensitivity of a

Varian detector that can be achieved by careful temperature control. Some of the design details that they discuss have now been incorporated into commercial versions.

Solutions have also been sought in changing the flow pattern[111] bringing the flow in at the middle so that half the flow goes towards the source and the other half towards the detector: the superiority of this configuration has not been confirmed unequivocally by other workers. A recent tapered cell[161] revives an old idea to reduce reflections but sacrifices cell volume for lower noise, an expensive and retrograde step when existing cells are already rather large (and other manufacturers achieve similar results with parallel cells).

Meaningful testing of flow sensitivity is not easy. Some confusing variables (many of which are difficult to control) are: changes in cell temperature; heat generated by the work done in forcing liquid through the column; thermal conductivity, mass and surface roughness of the cell; length, bore and material of connecting tubes; thermal conductivity and capacity, and temperature coefficient of refractive index of the liquid. The latter factor makes data from water optimistic when predicting results with organic solvents whose coefficient is several times higher.

Emission line plus filter systems

summary • **wavelength stable and narrow bandwidth**

Less commonly, other lines from mercury and other element lamps are selected with filters to provide a choice, albeit limited, of wavelengths. The advantage of filter wavelength selection is that it is extremely frequency stable and after effective filtration should have a very narrow bandwidth with low stray light. No wavelength recalibration is necessary or possible and no change in detector sensitivity can come from this source. Nor is there any problem with resetting. These advantages combined with the low cost and high stability of the mercury lamp account for its popularity. It can give the lowest noise in absolute absorbance terms, though not by a very significant margin (although sometimes detectors at other wavelengths may be much more sensitive in terms of concentration).

A variant of the 254 nm type is to use this light to cause a phosphor to emit light at 280 nm with a half bandwidth of 20 nm, very much broader[145] than the parent line. Simple mechanical changes can allow this changeover and some configurations allow simultaneous 254 and 280 nm detection. This is useful for some biochemical analyses, especially combined with the sum and or difference signals. However the more interesting wavelengths still lie below 254 nm and the steep rise of absorptivity with wavelength make a continuously variable system rather than widely spaced emission lines almost essential to work below 250 nm.

Low cost of 254 nm types arises from the cheap and reliable source with ample light output, and from the simple photodetectors that can be used. Photodetectors are usually (at least in the cheaper models based on the Watson patent[78]) a visible light-sensitive solid-state photodiode detecting light emitted by a phosphor proportional to the UV light (254 or 280 nm) reaching it. The system is double beam in that the lamp output is monitored and it may be possible to pass liquid through the reference cell. As noted below this is rarely as attractive in practice as in theory.

Transmittance to Absorbance Conversion

summary
- **absorbance linear with concentration**
- **transmittance gives log (concentration) display**
- **error in zero transmission setting lead to non-linearity**

Absorbance output is essential because it is linearly related to concentration. Transmittance out is also useful and underused: it gives a log (concentration) display that avoids range changing to keep peaks on scale and is ideal for exploratory work.

Electronic logarithmic amplifiers can now make the conversion with an accuracy exceeding that of other elements of the system. To generate the function log (I/I_0) it is possible to either compute the ratio I/I_0 and take the log, or evaluate $\log(I)-\log(I_0)$. The latter seems to give higher accuracy. In the case of single beam instruments $\log(I_0)$ is taken as an arbitrary reference voltage and I adjusted by optical attention of the beam to set zero absorbance. This avoids the need to match log amplifiers precisely and it is easier to obtain an accurate conversion.

A check on the conversion accuracy is easily made on instruments displaying both transmittance and absorbance by comparing the readings with calculated values (for example section 11.4 or reference 79). Calibration against sample concentration checks all types for accuracy and linearity but is time consuming. Published data proving the accuracy of photometric detectors is sparse.

A common source of non-linearity at high absorbance is failure to set the zero light or 0% transmission level correctly. This control, which may not be accessible to the user, sets the electronics such that zero light reaching the detector corresponds to zero transmittance. This sets the amplifiers to try to produce an infinite absorbance reading when no light passes through the cell. Light leaks can also upset this setting; shining an electric torch at all joints in the instrument will usually trace any leaks which might otherwise only show up when the sun chances to move round to fall on the detector.

Variable Wavelength Detectors

summary
- **varying wavelength can increase or decrease sensitivity**
- **selectivity more important than sensitivity gains**
- **single beam type best unless scanning sought**

Continuously variable wavelength selection requires a continuum source and a monochromator. The advantages of this facility are now well recognised (though it is interesting to note marked differences in usage between Europe and the USA: two commercial models were available and widely used in Europe some time before they became popular in the USA). The usefulness of wavelength selection varies widely from nil to absolute necessity depending on the compound(s) being analysed. It is not just extra sensitivity that can be achieved: it is also less sensitivity, useful for preparative work where the sample size cannot sensibly be reduced. Even more important is the way in which selectivity can be changed and this is especially relevant because of the low linear range of photometric detectors ($\sim 10^4$) compared

Changing the detection wavelength

Column: Spherisorb S5W, 100 x 5 mm bore
Sample: 3-isopropylphenol (Aldrich tech grade no 1-2658)
Eluent: hexane:ethanol (99.5:0.5 v/v) (Rathburn Chemicals Ltd)
Detection: Cecil Instruments CE212 at wavelengths indicated

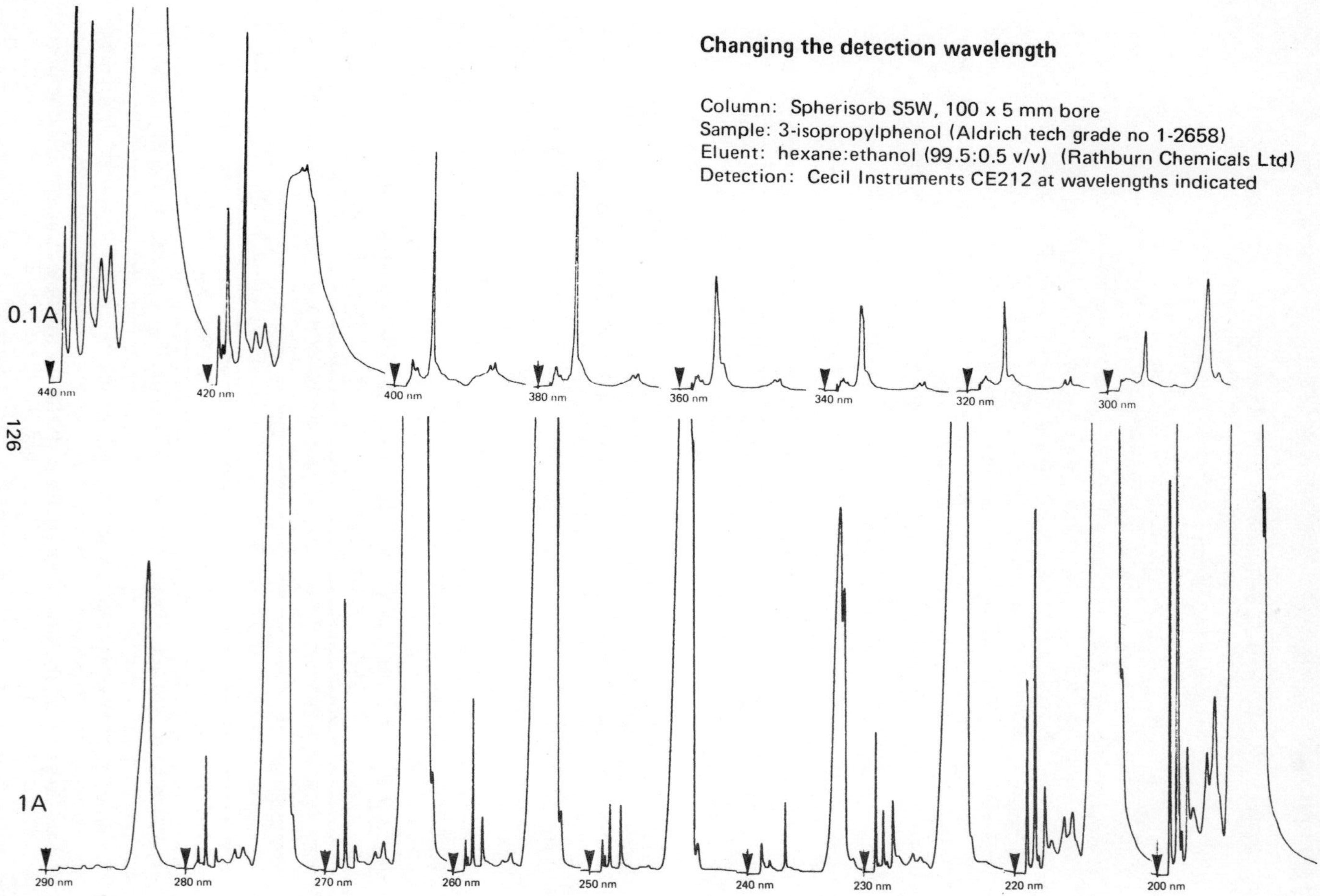

to the FID of GC ($\sim 10^6$). This limited range is dictated by the 'antilog' response of an absorbance system. If the ratio of two compounds present in widely differing concentrations is to be measured it is desirable to choose a wavelength where absorbances of the two peaks is as near equal as possible. Additionally, selectivity can substitute a little for separation, avoiding for example, a small shoulder on a massive peak. Although it is an extra parameter to vary, wavelength selection can be important in practice.

A single injection detected at two different wavelengths can give quicker results and a more accurate ratio than two injections into one detector set in succession at two wavelengths. A dual detector can be useful but the common wavelengths of 254 and 280 nm are rarely the optimum. Two detectors can be connected in series, but the elaborate architecture surrounding some chromatographs makes this type of link difficult in practice, which is a pity.

Variable wavelength monitoring has associated disadvantages, of course. Demands on the source stability in both single and double beam types are stringent. Lamp noise seems to arise mainly from arc position instability. This is incompletely compensated by using a double beam system because the amount of light passing sample and reference optical paths is affected differently by changes in the source location. Double beam systems may also introduce noise from optical and/or electrical chopping systems. In view of the simplicity and cheapness of the single beam system, it seems the better compromise for an LC detector unless a scanning system that gives spectra is sought.

At any selected wavelength, the light output of a continuum source like the deuterium lamp is small compared to a line spectrum emission lamp like the mercury lamp where most energy is concentrated at a few wavelengths. This means that more sensitive photodetectors are essential, either photocells, or even photomultipliers. Increasing bandwidth by increasing monochromator slit width also becomes a sensible compromise, as does an increase in flowcell aperture, or a more efficient optical system design to persuade more light to pass down the cell and emerge within the collection angle of the photodetector. Because most absorption bands in LC are quite broad, a wider bandpass has little effect on the results obtained. A nominal 15 nm bandpass (an ill-defined value at wide slit widths and to be compared with other manufacturers' values with caution) is a reasonable compromise.

Further increases lead to loss of sensitivity and linearity at high absorbances when measurements are made on the side of a band. It is easy to predict and avoid problems from wide bandpass by considering its effect on a spectrum of the compound concerned. Naturally measurement at the peak maximum reduces these losses.

Wavelength selection accuracy and precision is lower than with fixed wavelength types. Though not usually serious, the effect of slight changes in wavelength must be considered. Measurement at peak wavelength again tends to reduce sensitivity to wavelength setting accuracy. For the highest precision it is important not to alter the wavelength setting between chromatographing standard and sample. A detent mechanism, to give click stop selection every nanometre or so, is another way of improving wavelength setting precision. Its nominal setting accuracy of 1 nm may be combined with a resetting precision of better than 0.1 nm (which only alters if the instrument is realigned by a service engineer). In practice there are so many other sources of variation in a complete chromatographic analysis that when high long-term accuracy is required calibration with standard solutions and internal standards is probably also essential.

Double Beam Photometer Detectors

summary
- **double beam less useful than expected**
- **convenient for prep where wavelength controls sensitivity**

In theory a reference cell should be able to correct for changes in eluent absorptivity: in practice it is too difficult to match two streams precisely enough and it is better to accept a smooth change in the baseline and remove its effects by data processing by hand, or using a computer or integrator. This has the added advantage that it allows some check on the eluent composition profile. Matching the impedance of two columns, one for separating feeding the sample flowcell, and another supplying the reference beam flowcell, is just feasible, especially with slurry packed columns, but injection of real samples containing residues and septum debris tends to upset the initially balanced flows.

In practice a double beam system is only really useful (and value for money) if it can produce spectra which can help with *sample identification*. With a monochromator type it is certainly also convenient to be able to change wavelength without the serious baseline shift that would occur on a single beam type because the lamp output varies with wavelength. This is especially so for preparative work where wavelength usefully controls sensitivity.

Scanning UV-VIS Detectors

summary
- **UV spectra with flow stopped now feasible with some spectrometers**
- **baseline flatness at high sensitivity key problem**
- **high-speed, continuously scanning, computer based, on-the-fly detectors being developed**

UV spectra provide a limited amount of identification information. If UV detection is used, spectra can be available with little extra effort, and at modest extra cost. Scanning double-beam UV (and perhaps visible) detectors are needed to record spectra. Those now commercially available are more (or in most cases, less) developed from existing designs for UV-VIS spectrometers and scan so slowly that flow must be stopped during scanning. Because no certain market for this type of detector has existed little development work has been evident from manufacturers products which meets the rather difficult requirements of chromatographers. In the past it has always been possible to collect a peak and use a conventional microcell in a routine spectrometer; this is no longer possible because efficiently packed microparticles yield micro-volume peaks which are far too small for transfer except using the pipework and flowcells used in LC detectors.

Compared to routine spectrometry applications there are four problem areas:

1 Micro flowcells require small apertures and long pathlengths.
2 High sensitivity to small absorbance changes demands very low noise.
3 Useful spectra need good baseline flatness even at high sensitivity and with micro flowcells.
4 Scan speeds are far too slow to record spectra 'on-the-fly'.

Small cell apertures require efficient light utilization, partly met by the best types which have beam geometry that focusses the source image to a few millimetre square and pass $>$10% of the available light. It is difficult to radically increase the sensitivity of the photomultiplier, or electronics, or the lamp output, so it is important not to waste too much light. A fewfold improvement over current designs may reasonably be sought here.

Beam balance is seriously upset by microcells because of the many total internal reflections and the loss of much of the beam. The effects are minimised by optimisation of the optical balance so that sample and reference paths are exactly equivalent. This requires some critical mechanical adjustments, and care in selecting windows for equal transmission. However it is still hardly practical to attempt to match to $<$0.5% transmittance (a sensible but not ultimate target for baseline flatness for LC use) and one must look to a beam balance adjustment mechanism for final trimming. The demands of LC are much greater than of straightforward UV spectrometry because sample size, concentration and pathlength are all fixed by chromatography and cannot be adjusted to give a suitable absorbance around 0.3 as is usual in spectrometry. First generation instruments[83] approach this target of baselines flat to 0.001 A but many are so far off that they are little use for drawing spectra unless a peak is greater than absorbance 1. In practice it is always the spectra of small impurity peaks that are of real interest. Second generation instruments will meet the 0.001 A baseline flatness target.

A third generation of devices is emerging which scan much more rapidly either by moving the source with a cathode ray tube, a system described some years ago[80], or an oscillating mirror[84], or by using electronically scanned multi-element detectors still under development where each element detects a small wavelength range[81].

Encouraging results have been described[84, 101], but the sensitivity in terms of signal to noise is not yet quite as good as for conventional UV detectors. Processing the data in more sophisticated ways will probably improve this; improvements to the quite new UV sensitive linear photosensitive arrays can also be expected.

An important extension to the *selectivity* of the photometric detection principle is demonstrated by Milano, Lam and Grushka who use on-line computing to display first derivative spectra.

A small computer is essential to process the data like that of GC-mass spectrometry systems. Spectra or chromatograms at any number of wavelengths simultaneously will be available. Beam balance correction can be made by subtracting the stored solvent spectrum. The remaining problem for an LC detector is the high accuracy of light intensity measurement needed (1 in 10^5) at high speed to give sensitivity comparable to that of existing detectors. To some extent these problems may be solved by data processing methods based on signal averaging, but in the long term collecting more data from multiple detectors is likely to prove superior.

The Scanning UV Detector for Identifying Peaks

Column: Spherisorb S5W, 100 x 5 mm bore
Sample: 3-isopropylphenol (Aldrich tech grade no 1-2658)
Eluent: hexane:ethanol (99.5:0.5 v/v) (Rathburn Chemicals Ltd)
Detection: Cecil Instruments CE515

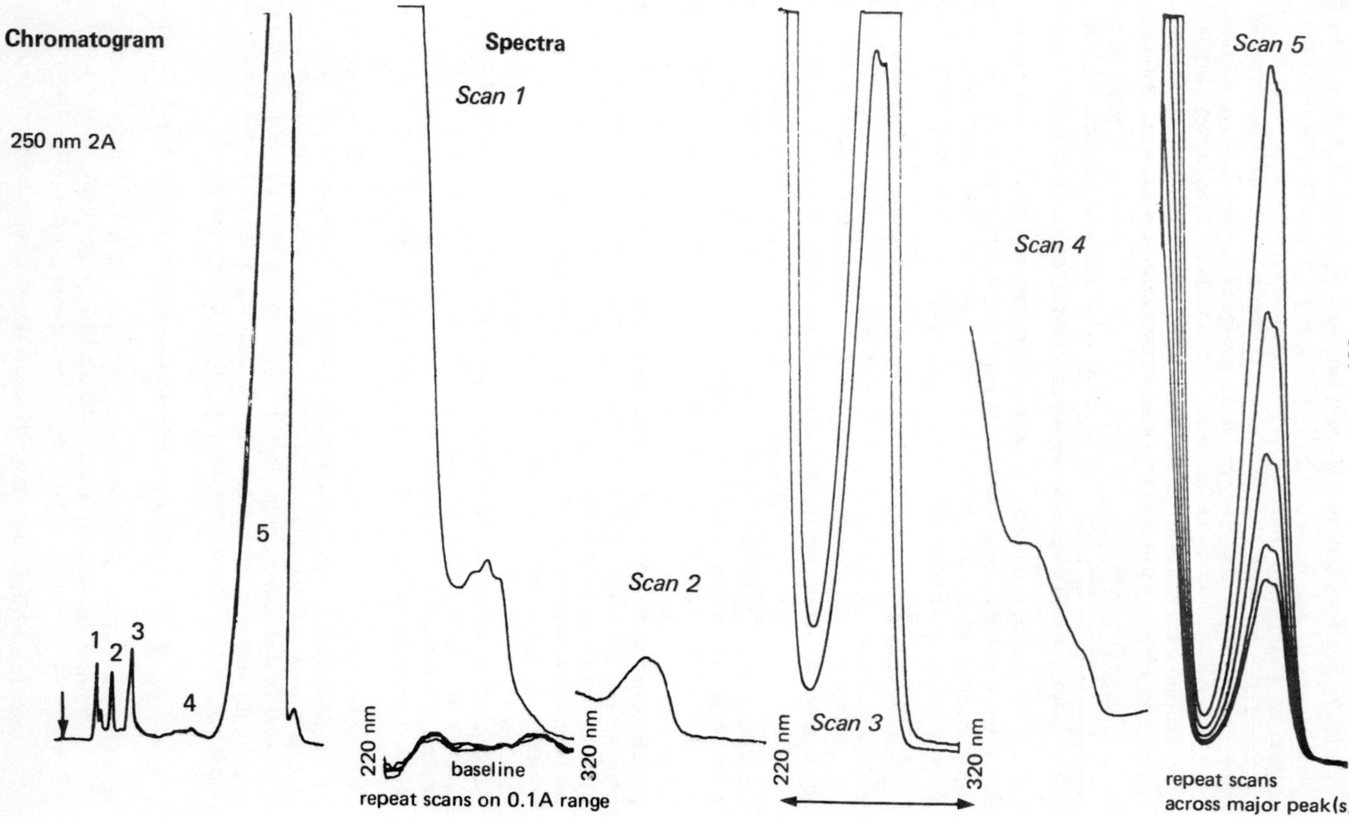

Infrared Photometry

summary
- **no transparent polar solvents for adsorption chromatography**
- **most useful for permeation chromatography**
- **IR spectra need solutes recovered from solvent**
- **multi wavelength IR potentially powerful for copolymer analysis**

As noted above, the main difficulty with the infrared (IR) region for LC detection and identification is the relatively small difference in absorptivity between solute and solvent, especially those solvents suitable for adsorption chromatography. For monitoring some of the bands listed in the table below there is no problem in finding relatively non-polar solvents, but there are almost no suitable high polarity solvents apart from prohibitively expensive deuterated solvents. This limits application of adsorption chromatography to fairly non-polar compounds.

Some Infrared bands for LC detection

bond	mode	wavelength μm	typical absorptivity mol/(l cm)	possible solvents
O—H	saturated stretch	3.0	50—200	most halocarbons, 1,2,3-trichlorobenzene (TCB)
C—H		3.4	50—200	CCl_4, C_2Cl_4, C_2HCl_3, CS_2, CH_2Cl_2, CCl_3CF_3, *o*-dichlorobenzene
N=C=O		4.5	2000	CCl_4, C_2Cl_4, $CHCl_3$
C=O		5.8	300—700	most halocarbons
Si—O—Si	bend	9.3		CCl_4, *o*-dichlorobenzene, C_2Cl_4, CH_2Cl_2
C—H aromatic	out-of-plane deformation	14.35		C_2Cl_4, 2,2,4-trimethylpentane, cyclohexane, *o*-dichlorobenzene, $CHCl_3$

With permeation chromatography, however, several factors have led to its successful use: many polymers are not detected by UV photometry, for example polyalkanes and polyolefins; chlorinated solvents, which are already widely used, are transparent in the C—H stretch region; permissible cell volumes are larger than usual with adsorption chromatography; and a lower sensitivity is needed because sample loading is high and dilution small.

Several hardware implementation approaches have been made. Some conventional double-beam scanning IR spectrometers have been adapted for monitoring. As with UV the requirements for an LC detector are slightly different from those of a scanning spectrometer: low volume flowcells; ability to operate at the high temperatures needed to keep solutes in solution (up to 165°C); high zero absorbance stability, and detection of very small changes in transmittance with low noise and drift; and if scanning is intended, double-beam operation with a small baseline disturbance caused by small aperture cells. Wavelength selection may be by monochromator or a filter. Instruments that have been adapted are the Grubb-Parsons and the Wilks Miran interference filter instrument. Freeman et al[82] describe the use of an infrared laser (He—N at 3.39 μm) to monitor cholesteryl palmitate peaks eluted from a pellicular silica column with C_2HCl_3. The use of a multi-channel IR process analyser has also been proposed[169]

Flowcells are usually only 3 mm pathlength. Window materials that have been used are: polythene, PTFE, alumina, silica (limited range) and sodium chloride (rather easily damaged). At wavelengths >7 μm alumina is most suitable (see section 10.4).

IR spectra contain considerable identification information. Because of the poor transparency of solvents it is hardly practical to obtain a complete scan except by collecting the peak, evaporating the solvent and running a spectrum of the residue. By monitoring successively two different functional groups the relative amounts of two components of a copolymer have been measured by Hemming and Dawkins[172]. For good accuracy they found an internal standard was needed. With a multi-wavelength instrument it will be possible to monitor both groups simultaneously and obtain the ratio information from one injection.

Fluorometric detection

summary
- **very selective and sometimes very sensitive**
- **absorption of light causes non-linearity and needs monitoring**
- **fluorescence spectra can identify peaks**
- **pre or post column derivatisation useful**
- **large diameter columns needed for highest sensitivity**

Selectivity is the main attraction of fluorimetry. Sensitivity may also be very high in suitable cases.

Derivatisation with fluorescent labels extends this detection method to the very many compounds that have no natural fluorescence. The sample to be analysed may be reacted with the derivatising reagent and the labelled compounds chromatographed; or the column effluent can be mixed with a reagent and the resulting solution passed, perhaps via a heated reactor section, to the fluorimeter. Fluorescamine is a typical reagent for forming derivatives of primary amines[170] and has been used for both pre- and post-column derivatisation. A post-column reaction based on the production of fluorescent cerium (III) has been used by Katz and Pitt[171]. The column effluent is mixed with a solution of cerium (IV) and any compounds that are oxidized under the reaction conditions produce cerium (III).

To detect fluorescence a UV source excites the sample which re-emits light at a longer wavelength. The source may emit discrete lines, such as a mercury lamp, or a continuum like the xenon lamp. The emission line or wavelength bandpass[139] must be selected by a filter or monochromator. As with UV spectrometry, filters have a fixed bandpass and are cheaper[194], but monochromators are more flexible. Filters may also have smaller stray light transmission (light at unwanted wavelength) than monochromators.

Emitted light is usually viewed at right angles to the exciting beam and a filter or monochromator used to block scattered light at the excitation wavelength. Background fluorescence and scattered light, both zero offsets for LC monitoring, are tolerable provided not too large and constant. Absorption of the exciting (or emitted) light is a potential pitfall: if the exciting light is absorbed by the solution it cannot excite the whole volume of the flowcell and less light will be emitted and the response non-linear. Light may be absorbed by the fluorescing compound, or any other compound eluting at the same time. This is a potentially serious practical complication. It is therefore prudent to monitor UV absorption as well, to guard against this: absorption should not exceed 0.1A for reasonable linearity. The Dupont instrument[192] provides this information as an integral part of the fluorimetric detector: it may be more flexible and economical to couple a variable wavelength UV detector with a fluorescence detector.

Suitable flowcells for HPLC which fit into existing fluorimeters are now becoming available. The criteria for a flowcell are rather different from those for absorption detectors. The emitted light, and thus the sensitivity, depends on the intensity of exciting light (which is controlled by the source power and focussing efficiency) and the cell volume irradiated. Thus the chromatographic requirement for low peak broadening conflicts with the detector volume requirement for sensitivity. This can only be resolved, for the highest sensitivity, by using a larger volume flowcell but increasing the peak volume to compensate. Thus instead of an 8 μl flowcell being used with 2 mm bore columns, a 45 μl flowcell (1.5 x 1.5 x 20 mm) can be used with a large diameter column, say 8 mm bore, giving peaks that are 7 times the volume and can be detected without too much resolution loss. A long thin cell will give less zone spreading than a short fat one of the same volume (see section 12.4), and be less sensitive to quenching effects. Unlike UV flowcells, there is no need to converge the light through a small aperture; a long slit is preferable. Specifications on sensitivity need to include the flowcell to be used: those on the basic instrument with 10 mm square cuvettes are not necessarily relevant.

Instruments designed specifically for UV detectors are usually simpler and may be cheaper, but have less flexibility with little or no choice of wavelength selection. Unless a derivatizing agent or compound of well known fluorescence spectrum is always being used, this may prove a serious limitation, particularly to the degree of selectivity that can be achieved.

A spectrum obtained by scanning the excitation or emission wavelengths (or both) can give considerable identification information[194]. To obtain this it is necessary to have an instrument with one or two monochromators (or variable wavelength interference filter) which can be scanned.

7.6.2 Refractometers

summary
- **RI troublesome and insensitive for analytical work and second choice to photometric detectors**
- **widely used in permeation chromatography**
- **more useful in prep LC**
- **all types very sensitive to temperature, pressure and flow**

Refractive index (RI) measurement has always been used in permeation chromatography where sample concentrations are quite high. It has never become very popular in other types of liquid chromatography because of its low sensitivity to solute and high sensitivity to temperature and pressure. Design improvements have reduced but not removed these handicaps and the higher sample concentrations available from more efficient columns and higher loading used in preparative work suggest a slight renaissance for this type of detector.

Its greater universality can be helpful: UV detectors can be extremely misleading for indicating the weight of a peak. In practice it is still barely sensitive enough and is mainly used when UV detection fails, or in permeation chromatography where high sensitivity is less necessary. (High sensitivity is not redundant in permeation chromatography[85].)

Even in permeation chromatography of polymers, the response is not an absolute one; molecules of different chain length or branching have different effective refractive indexes. Chromatograms need interpreting and comparing with caution.

A different range of solvents can be used because UV transparency is no longer necessary. Though this allows use of some solvents traditional in column chromatography it is not yet clear that these are essential except for those solvents needed to dissolve high molecular weight polymers for determination of molecular size distribution. Because sensitivity depends on the refractive index *difference* between solute and solvent, intelligent solvent selection can vary effective sample sensitivity severalfold.

Three basic types of refractometer are available commercially. A number of related detectors measuring physical properties like dielectric constant linked with refractive index have also been described. None of these have yet demonstrated a clear advantage technically or commercially over the more direct optical systems. All suffer from the same basic problem of temperature and pressure dependence which limits their usefulness for LC detection.

Fresnel

Fresnel's law relates the light fraction reflected at a glass-liquid interface to the refractive index and angle of incidence. For maximum sensitivity the angle is set to just retain total internal reflection in a prism. A reference beam, preferably with a reference flow, helps compensate for temperature and flow changes, and variation in output of the source lamp, usually a tungsten filament. The high output stability and long life of light emitting diodes[86] suggests they might be useful. The cell volume can be made very small, down to a few μl because the cell can be very thin, perhaps 100-200 μm, the thickness of a gasket. Larger versions which are more suitable for preparative flowrates can also be made easily. A sensitivity corresponding to a change of 10^{-7} in refractive index can be achieved, but with difficulty.

Deflection

A slightly different design depends on the deflection of light passing in turn through reference and sample liquids, caused by differences in their refractive indexes. The deflection is sensed by interposing a mask and measuring the light intensity reduction as the beam is deflected and is obscured by the mask. Although it is easier to obtain a high linear range, it is more difficult to design a small and cleanly swept flowcell than with the thin film of liquid needed for the Fresnel type.

Interferometer

Direct comparison of the velocity of light through sample and reference beams can be made with an interferometer. A commercial example is the Multiref 901 by Optilab. Birefringent elements first split and then recombine beams. Interference of the two beam wave fronts is changed by changing velocity of sample and reference beams. It is easier to produce a design with good stability (though perhaps at a higher manufacturing cost) but a stable flow and temperature are still essential. The linear range is limited by the way in which out-of-phase signals eventually come back into phase. The sensitivity range can be altered conveniently by changing flowcell pathlength. Bakken and Stenberg[159] describe a related interferometric system.

All these types, including those which measure related physical properties, depend critically on careful control of temperature and flowrate or pressure. It is particularly important to equilibrate the temperature and flow and pressure of the sample and reference liquids. A temperature difference of 10^{-4} K or 70 Pa cause most liquids to change RI by 10^{-8}. The pressure change corresponds to only about 10 mm of solvent head. Flowrate changes and pulsing flows produce noise from changes in thermal equilibration and changes in the pressure drop across the cell. Circulating water provides the highest heat transfer rates and it may be useful to circulate the same liquid round a column jacket to keep the whole system in thermal equilibrium, but is regarded as inconvenient by many chromatographers. Coaxial tubes containing sample (usually centre) and reference (outer) equalise the liquid temperatures well. Thermal conductivity of stainless steel is rather low. Too long or too large diameter tubing may cause zone broadening (see section 12.4).

In practice it is often troublesome to thermostat and equilibrate well enough to achieve the ultimate specification noise and drift levels.

REFRACTOMETERS

Fresnel

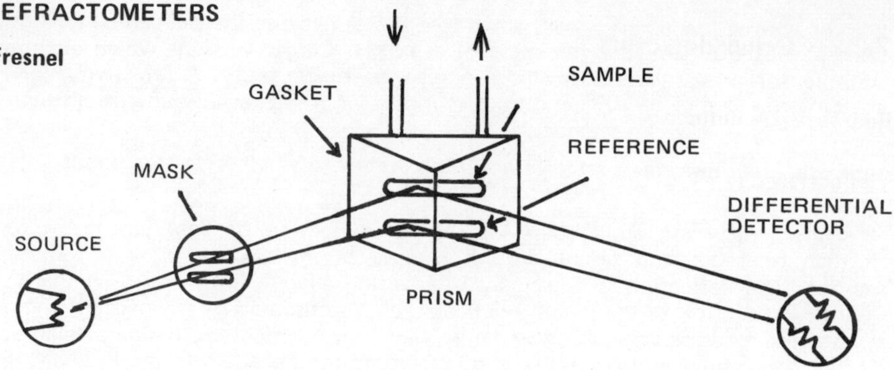

Deflection

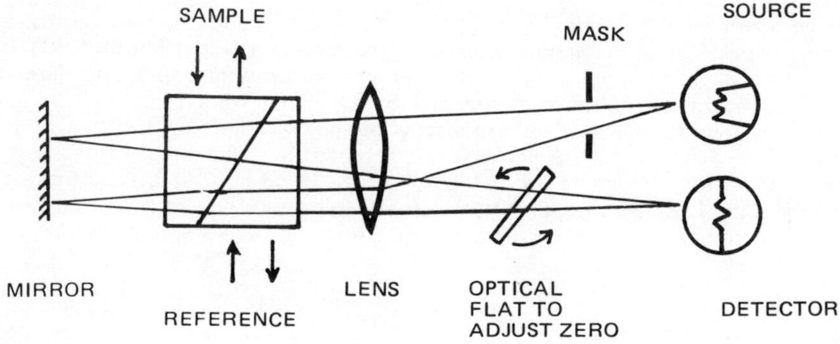

Interferometer

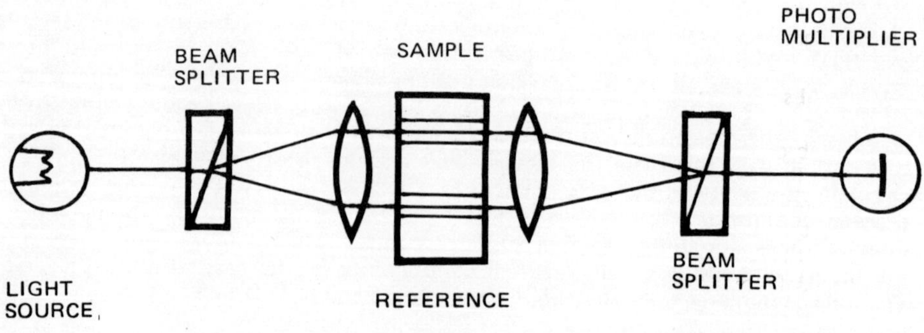

7.6.3 Other detectors

Radioactivity detectors

summary • **important detector not yet matched to very efficient columns**

In view of the importance of radioactively labelled compounds in many fields, especially pharmaceuticals, further development of this detector is important. Unlike detection of radioactive compounds on TLC plates, where the peak positions are frozen after development and can be detected at leisure, LC is a dynamic process and compounds must be detected in real time. Unless the specific activity of the sample injected is high, or the sensitivity of scintillation detectors is improved, the system must run too slowly for optimal chromatography with microparticles. Peak volumes are rather low for current flowcell designs[91] (>100 μl) and this can only be increased at the expense of observed column efficiency (which also reduces sample concentration) unless larger diameter columns are used. This implies more total activity on the column.

Another possible escape from this technological impasse is to stop flow, when peaks are almost frozen in the column and flowcell, and can be held without resolution loss, thus allowing long counting times. Stop-start chromatography may allow one to run at, say, 1/50 the normal minimum reduced velocity without the serious resolution loss that would occur with continuous flow at this speed. A further alternative is to collect peaks and examine these small volumes later by conventional methods. This method is found effective but tedious[140,225]

A number of scintillators are in use. These emit light as they absorb radiation from the solute. The light pulses are measured by a photomultiplier and their rate (and perhaps type) recorded. Some difficulties of compatibility of chromatographic eluent and scintillator exist. Efficiency of scintillators varies from about 5–70%. Sieswerda and Polak[94] conclude POPOP 2,2′-*p*-phenylene-*bis*-(5-phenyloxazole) is the most suitable for an HPLC flowcell and describe a 100 μl design. Schutte[155] reviews various possible scintillators, both homo- and heterogeneous. A number of other factors affect the performance of this type of detector: changes in optical absorbance of liquid in the cell, contamination and quenching, chemiluminescence and background radiation.

Atomic Absorption Spectrometry as an LC detector

Atomic absorption spectrometry (AAS) has proved a highly selective, and sensitive, detector for a restricted number of compounds containing suitable elements, mainly organometallics[98]. It is made practical by the ability of the flame to consume all the eluent and so avoid the difficult solvent-solute separation. This places limits on the solvents that are suitable, and on feasible flowrates for existing burner designs. Unlike UV detectors this type is mass not concentration sensitive: if the flow is stopped the signal will fall to zero as absorbing atoms are swept out of the flame zone. Jones et al[98] discuss the interesting variation of sensitivity with flowrate, solvent and element. Burner design specifically for LC is important and further development is expected.

This type of detector promises to make LC a rather important analytical method for certain classes of compounds because many, organometallics in particular, are only stable in solution at low temperatures. Photometric detectors can, of course, be placed before the AAS detector when the total information from the two detectors is likely to be more than just the sum of the parts.

Phosphorus and Nitrogen detectors

Julin et al[160] describe another detector that consumes all the eluent, passing it directly into a flame where emission from phosphorus at 526 nm and sulphur at 383 nm give selective and sensitive detection of compounds containing these elements. Wavelengths are selected with narrow bandpass interference filters and the light passing detected with a photomultiplier.

As with AAS the way the sample enters the flame is critical and the authors predict improvements from more sophisticated nebulizer and burner design. Only aqueous phases have been used successfully so far.

Separation and detection of inorganic species like H_3PO, H_3PO_4, H_3PO_2 suggest LC as an analytical method where it would not normally be used today.

Electron capture

It was clear to Willmott and Dolphin that to use the high sensitivity of electron capture effectively, it was essential to use all the liquid from the column, without the wasteful and complex solute transport system. They have shown how this principle can be applied to organochlorine pesticides[93]. The range of application is now being extended to some nitrogen containing pharmaceuticals and organophosphorus compounds. Some rather more polar eluent mixtures can also be used[190] provided strongly electron capturing compounds (as eluent components or impurities) and dissolved oxygen are excluded by sparging the solvent reservoir with nitrogen. Water and acetonitrile seem to be unusable even at concentrations $<1\%$. A commercial instrument is now available from Pye Unicam Ltd, Cambridge, UK.

Electrochemical detectors

Conductivity has not proved useful as a practical detector (though its very low volume and zone broadening commend it for column research—Pecsok and Saunders[144] describe a suitable design). In practice the conductivity of a chromatographically useful eluent is always large and the additional conductivity of the sample too small for a useful sensitivity.

Polarographic detectors using a variety of electrodes[162,164] have been produced (and progress reviewed recently by Veening[143]) but none are yet robust enough to have become popular with chromatographers. Commercial versions are now available[76,156] which will enable chromatographers to assess the extent to which these difficulties can be overcome, and the range of compounds for which this has advantages over other detectors.

The dropping mercury electrode used in the pioneering work of Huber[90] is excellent electrochemically but troublesome under high local flow velocities essential to avoid zone spreading in LC. Carbon electrodes are favoured[76,88,89,156] but there are problems with poisoning of the surface by some compounds and eluents. Other promising work[147], fails to meet the small flowcell volume requirements to allow modern microparticle-packed columns to be used. Lankelma and Poppe[229] favour a coulometric approach where all molecules react on a carbon electrode of large surface area (80 x 7 mm), but small volume because the liquid path is so thin (~50 μm). They claim this is predictably mass sensitive and partial deactivation or poisoning is less serious than with small electrodes used polarographically.

If all these problems can be overcome electrochemical detectors will prove very useful for they promise to provide a wide dynamic range, selectivity and some identification information as well.

Heat of adsorption detectors

Early in the development of high efficiency LC a number of detectors were based on the heat of adsorption of the sample onto charcoal or other active surface. These did not prove popular and are now virtually extinct.

Solute—Eluent Separation Systems

summary

- **solvent—solute separation (or transport) still not good enough**
- **key to using FID, and other specific gas-phase detectors and mass spectrometry**

Separation of the solute from the eluent and its transport is the key to using the flame ionisation detector (FID) and similar gas-phase detectors, and mass spectrometry. Despite many efforts by many workers over many years the separation and transport efficiency are still disappointing. The moving wire of Scott[149] is of longest standing and many units have been sold. It still fails to transport ~95% of the sample—and still transports too much residual solvent leading to noise in the detector. Even the outstanding sensitivity of the FID or mass spectrometer cannot tolerate this loss *and* the solvent residue signal without serious performance degradation. Its mechanical complexity leads to unreliability and its hot zones are a fire hazard. Modification of the wire surface and the methane conversion system[150] have been welcome improvements but the overall system is still not as effective as one might hope. Response speed is now too slow for microparticle columns, and the detector variance too large for small peaks from short efficient columns.

Several other transport devices based on woven meshes, chains, metal strips have been described but all show too much noise, mainly as spikes, and too low sensitivity. The most promising new idea has been a porous alumina disc which prevents sample diffusion by holding the material in the pores giving low noise and transporting a large fraction of the sample[151]. A commercial device that solves all these problems is needed to use all the proven gas-phase detectors now used for GC. Moreover, it is the key to that most powerful combination — LC—MS.

Liquid chromatography—Mass spectrometry (LC—MS)

summary

- **existing systems do not utilize the full identifying power of MS and are rather expensive to be just a detector**
- **solute—solvent separation key problem still to be solved**
- **discrete transfer after collection and solvent evaporation gives most advantages at present**

This technique is known to be capable of giving highly specific and sensitive detection with detailed identification information. The success of GC—MS, particularly with computer data systems, and the practical results that have been obtained by collecting peaks from analytical LC columns, prove the great potential of LC—MS but there are some technological hurdles to be overcome yet.

Wide analytical columns (1/4 in diameter, 5 mm bore or 3/8, up to 8 mm bore) have sufficient capacity to allow enough material, even of an impurity peak, to be collected for subsequent mass spectrometric identification. The solvent can be evaporated and the sample inserted directly on the mass spectrometer probe, or a concentrated sample injected into a GC—MS system where the function of the GC is solely to separate sample from residual solvent and decomposition products.

This is likely to remain a powerful peak identification tool because it is simple and cheap. (LC—MS is not going to be cheap.) Compared to GC collection it is very easy and almost complete recovery is usual. A stream of nitrogen or vacuum, perhaps with heating, can remove the solvent. Purity of the solvent is critical (see section 12.5). This discrete separation and on-probe injection into the mass spectrometer has been automated by Lovins et al[218]. The chromatograph is operated in an intermittent-flow mode to identify each peak. However these techniques cannot give chromatograms, nor the extremely selective ion-monitored method of detection.

One approach to the LC—MS problem, made by two groups led by Horning and McLafferty respectively, has been to accept the solvent and use it to form primary ions which react with solute molecules. Horning[154] uses an atmospheric source while McLafferty's system[153] is much more like conventional chemical ionisation mass spectrometry in only using a percent or so of the effluent. Both systems rely on low molecular weight solvent ions not interfering with the higher molecular weight compounds of interest. As developed so far neither system provides the extremely powerful identifying features of conventional electron impact MS, though the identification data is very useful. To achieve the goal of getting all the sample and very little eluent into the spectrometer so that all the range of methods can be used as with GC—MS, some breakthrough in solute—solvent separation is required. The problem is several orders more difficult than that of carrier—solute separation in GC—MS because the solvent and solute are now relatively much more similar. A silicon rubber separator has been described[152] but is limited to rather polar solvents and non-polar solutes.

McFadden, Schwartz and Evans[242] have recently described some progress with the LC—MS problem (which McFadden has characterised succinctly as 'damn hard'). Their apparatus transports all the column effluent on a loop of stainless steel

3.2 mm wide, 50 μm thick at a few cm/s through chambers with steadily decreasing pressure where solvent is pumped away. The belt finally passes into the mass spectrometer itself where it is heated to evaporate the solute. A very small clearance between belt and sapphire slits reduces loss of vacuum. About 30% of solute is detected by the spectrometer section but spray losses cause this to vary rather much for good chromatograms. The identifying power of MS is very clear nonetheless. Improvements and refinements are acknowledged as necessary for a practical commercial device but LC–MS is clearly feasible.

New detectors

There is still a need for novel detectors in LC. A number of descriptions of new detectors in the literature have yet to reach commercial model status which may indicate some practical snags: many still have promise. Karasek[92] has reviewed the state-of-the-art in 1975.

LC Detectors

Summary

1 **UV first choice** detector unless proven unsuitable in practice.

2 **Variable wavelength flexible**, often most **sensitive to sample,** especially at very short wavelengths, and **selective.**

3 254 nm types can have **best signal/noise** and linear range, and should be cheapest.

4 Double-beam only really worth extra cost for scanning to give spectra.

5 Beam balance and **flat baseline at high sensitivity** critical parameters for **scanning UV** detectors.

6 **RI** good for **permeation and prep,** but insensitive and troublesome for analytical.

7 **IR** useful for **permeation** but rather **insensitive and solvent limited** for other types of LC.

8 **Fluorometric** detectors (perhaps plus pre- or post column **derivatisation**) useful for **selectivity,** and sometimes high sensitivity.

9 Transport/solvent separation plus FID universal but slow, insensitive and poor operability.

10 **LC-MS potential power,** but collection of peak, solvent removal and conventional mass spectrometry of **recovered sample** best at present.

11 **Selective** detectors like AAS and EC invaluable for specific problems.

12 Other detectors needed, mainly for **selectivity** rather than universality, but development of **robust** and **reliable** instruments is vital for wide use by chromatographers.

7.7 Data Processing

summary
- **integrators and data processing equipment and algorithms designed for GC generally suitable**
- **negative peaks can confuse some systems**

Algorithms and instrumentation designed for GC are generally suitable but some problems do occur. Negative-going peaks, either from RI detectors; or from sample or solvent in UV, often a refractive index effect at the solvent front, cause confusion in algorithms catering only for flame ionisation detector signals, often setting erroneous baseline values. This difficulty must eventually be resolved by modifications to the algorithm by manufacturers.

To utilize existing equipment it is usually necessary to suspend data collection or computation during the time window when negative peaks will occur. This facility is usually available but it may then prove difficult to find a suitable flat zone for the integrator to take as baseline. RI detector output is more difficult to deal with by this sort of subterfuge. A solvent change to one with a different refractive index may ensure that all peaks are positive.

The range of signals produced by most LC detectors is only about $1:10^4$, certainly not the 10^6 to 10^7 which the FID can produce. Thus 12-bit (1:4 000), 14-bit (1:16 000) or even 16-bit (1:65 000) conventional analog-to-digital convertors can be used without any need for range changing essential to cover the 23-bit ($1:10^7$) variation in signal found with the FID.

Further details on the problem of quantitation in LC is given in section 8.4.

8 STRATEGIES

8.1 Developing a Separation

summary
- **still an art**
- **few separation systems are unique**
- **search the literature and other information sources**
- **TLC is verv useful survey method**

Chromatography is still an art: there is no algorithm that can be followed blindly to achieve separation. Yet it is vital. Amos and Perry[367], comparing TLC and HPLC, observe: *"No matter how sophisticated the instrument, HPLC can be a very disappointing technique if the correct chromatographic conditions are not being used."* Seasoned users will only quarrel with one word —*"the"*— for few separation systems are unique: because a separation is achieved by one group of workers or in one particular applications laboratory does not necessarily mean it is the only method, or the best, particularly if large particle packings have been used.

However it costs nothing to use an existing separation (unless it means buying expensive equipment for gradient elution which is not really essential) whereas developing a new separation needs skill, time, and thus money, and has a small chance of failure. Searching the literature, in its broadest sense to include all possible sources of information, must therefore always be the first step. An old-fashioned method may often be quickly translated to more modern columns, or to use more common or standard solvents (see section 12.5). Existing TLC systems are also a valuable guide on where to start with an LC system.

Paradoxically, a TLC separation, even if resolution is incomplete, is a very useful survey in preparation for developing a column method, especially if the mixture is largely unknown. The reason is simple: TLC reveals the position of all solutes, even if they have not moved from the baseline. With a column, only solutes that are eluted are detected.

Translating from TLC to Column chromatography

summary
- **translation may be straightforward—or very difficult**
- **less polar mixtures usually used for column chromatography**
- **try half concentration of more polar component**
- **non-equilibrium (de-mixing or secondary front) a serious complication**
- **solvents for HPLC detectors usually need to be UV transparent**

In practice translating separation systems from TLC to HPLC is not quite as straightforward as one might naively expect, though several studies have shown that there are no theoretical differences[368,369] (see also section 12.11).

The first difference is quite predictable. TLC displays the answers as R_f values: LC in column capacity factor k'. R_f 0.5 is about halfway up the plate, and is a reasonable degree of retention to aim for, but the corresponding k' of about 1 does not take much advantage of the potential for increasing resolution on a column system by eluting longer, or further. So it is usual to use less polar eluents for column chromatography. One might try using half the concentration of more polar liquid as a starting point when transferring from TLC to HPLC.

The second difference is much more complicated. It arises because TLC, like dry column chromatography[370] is a *non-equilibrium* technique. This means that the eluent composition varies with time and position on the plate. This can, and has been, exploited to provide some rather good separations that HPLC has difficulty in matching, but is a major barrier to simple transfer of eluent systems from TLC to HPLC.

Preferential adsorption of a more polar constituent of the solvent is particularly a problem when using mixtures of components with widely different polarities, for example alcohol—alkane mixtures. If a secondary front is apparent, problems can be anticipated if the same solvent is used in a column at equilibrium. The results of 'solvent demixing' is to deplete the eluent of strongly adsorbed component some distance behind the solvent front: thus a much smaller concentration of polar components, or a less polar component, may be indicated for a column in equilibrium with the eluent. The dryness or degree of activation, and the rate of equilbration with solvent vapour in the TLC tank also control the eluent composition gradient, and thus the separation.

It is rather like a built-in gradient elution system which can be adjusted by changing eluent, activity and dryness of plates, and the degree of equilibration with solvent vapour. Small wonder then that there are sometimes problems of reproducing separations in different laboratories using different silicas and different apparatus: standardising conditions in HPLC should be easier, provided gradient elution is used with discretion.

A final problem in translation is a practical one: the predilection of TLC workers for solvents which are UV opaque. The solvent list in section 12.5 may help in selecting a suitable replacement which allow a UV detector to be used.

In the absence of a foolproof set of rules guaranteeing separation, the answers to a few questions may provide guidance on what to try first. Inspired serendipity has solved some difficult separation problems: rationalisations have followed later.

Known or Unknown?

Separating unknown substances is obviously more difficult than even slightly known samples: but certain lines of attack are more promising. The information that other analytical techniques can give should not be spurned: the forte of LC is high resolving power and accurate quantitation. One might even go so far as to say it should not be used to analyse unknown samples. TLC with a variety of solvent systems is certainly a minimum exploratory step.

Detection

It is obviously important to find out if all, or only some, of the substances can be detected by UV, the usual detector for LC. Otherwise considerable effort may be expended studying the separation of a few impurities while remaining oblivious to the major constituents, and with the possibility of chromatographic complications from large injections of compounds which are invisible to the detector.

A spectrum of bulk sample or concentrated solution run in a conventional UV spectrometer will show whether *any* compounds will be detected (but not whether they *all* will).

Collecting the major peaks, removing solvent and weighing (comparing to the weight of a solvent residue blank) may give some clue as to whether major components are being eluted and detected successfully. Usually >1 mg must be injected.

Exploratory LC

Gradient elution with organic–water mixtures and non-polar packing materials, currently usually bonded-ODS (and perhaps carbon in the future), have proved popular for tackling largely unknown samples. The other approach, advocated by Scott[375], uses a multicomponent gradient and silica, a polar material, as adsorbent, and a fairly universally (in)sensitive detector.

Is it Volatile?

Although LC will separate volatile compounds just as well as involatile ones, it may still be easier to separate volatile mixtures by GC. However LC has much to recommend it for the analysis of *mixtures* of volatile *and* non-volatile compounds.

Size?

If the material has a molecular weight above a thousand or so then the distribution of molecular size, oligomers if not polymers, is likely to be more interesting than the functional groups. To collect this information (gel) permeation chromatography (section 6.4) with quite large pore packing material are used. Conventional adsorbents like silica also separate by size, and start to discriminate at about the same size. This may provide a way of allowing all large (and thus excluded) molecules to elute first, and then separate small molecules by adsorption chromatography which discriminates by functional group and other molecular properties.

If samples are water soluble, like most molecules of biological interest, then dextran or polyacrylamide gels, or silica, are usually used. If samples are organic soluble, like synthetic polymers, the polystyrene gel or large pore silicas are usually chosen.

What is it Soluble in?

Solubility is by far the most useful guide for selecting a suitable solvent system. It can be used whether the adsorbent is polar, like silica, or non-polar (or reversed-phase), like bonded-ODS. In general compounds are eluted by the solvents they are most soluble in.

Thus if a compound like a phenol is soluble in methanol, but less soluble in hexane, it will be adsorbed on silica, a polar surface, from solutions in hexane and eluted more quickly as more methanol is added to the eluent.

In reverse, an aromatic hydrocarbon which is soluble in octadecane, and thus by analogy in bonded octadecylsilane, but rather insoluble in water, will be adsorbed from a largely aqueous eluent and only eluted by increasing the concentration of a less polar constituent like acetonitrile, in which the hydrocarbon will be more soluble.

These over-simplified examples are only a guide on where to start. In practice, and in theory, separation is much more complicated. To predict some of the complicating features it may help to consider the local changes in eluent composition in the stationary zone in the pores caused by interaction with the surface of the pore walls. Polar surfaces will adsorb polar constituents, like the methanol from hexane: non-polar surfaces will adsorb non-polar ones, like acetonitrile in the reverse-phase example. Retention of solutes will partly be based on solubility in the liquids in these zones, particularly the end which is aligned away from the surface. So changes here may produce the desired change in selectivity.

The main adsorbents are silica and alumina. Differences between them, and the various types and makes, are quite minor. It is usually possible to transfer separations from one material to another with only minor changes in eluent composition, and minor changes in selectivity.

Popular solvent systems are alkane—alcohol, alkane—acetonitrile, with added medium polarity compounds like chloroform, for solutes for which these are good solvents; or alcohol or acetonitrile—water systems for highly-polar water-soluble compounds.

Non-polar materials are solid organic materials like porous polystyrene gels (not only useful as sizing materials), ion-exchanging resins (whose ion-exchanging ability may be irrelevant), and carbon; and bonded or non-bonded layers of organic material like bonded-ODS or non-bonded β,β'-oxydiproprionitrile for LLC.

Alcohol—water and acetonitrile—water systems are used almost exclusively except for ternary LLC systems (see section 6.2.1).

Acids and Bases

The effectiveness of many separations lies not in the major components, but in the 'special ingredients' at the few percent level. They may affect tailing, peak broadening and most important, selectivity. Usually acids and bases are important components of solvent systems. Often they are present without deliberate addition either from an intrinsic property of one component; or because a minor impurity or breakdown product is acidic or basic. The latter reason can result in chaotic changes in separation if the source or batch of solvent is changed. For this reason it is often prudent to add say 1% of acid or base. A useful rule of thumb is to add acid for acidic solutes, for example acetic acid for compounds with a carboxyl group; and bases for basic solutes, for example ammonia or diethylamine for nitrogen containing compounds. Salts which buffer or vary ionic strength and solvent polarity may also be useful. Sometimes a theoretical rationale for these practical actions can be conjured up.

Is it Ionic?

Many water soluble compounds are ionic and difficult to separate on adsorption chromatography without some special procedures. Ion-exchanging column packing materials provide a way of separating ions (see section 6.3). If ions are not to be separated, it is important to avoid them by forming ion pairs, or other more permanent derivatives. Otherwise one may be trying to chromatograph several species at once, say an acid, its ion, and ion-pair—often an enterprise doomed to failure. In practice this may be an unconscious and simple step like ensuring that acids stay protonated, or a conscious step like forming an ester of the acid (see section 8.10).

Aids to developing a Separation

known separation methods
- ● search literature and other information sources
- ● try TLC with several solvent systems

known properties of the sample
- ● other analytical techniques needed

detectability
- ● are any, all or only some components detectable by UV?

size (say mw $>$1000)
- ● is molecular size distribution of interest?
- ● is large-from-small fractionation useful?

if water soluble —
- ● try dextran or polyacrylamide gels, or silica

if organic soluble —
- ● try polystyrene (GPC) or silica (large pore)

solubility
- ● predict or assess solubility in various solvents

if soluble in non-polar organics —
- ● try alkane plus polar component on a silica microparticle column
- ▲ vary concentration of polar component doubling or halving concentration at first
- ▲ avoid $<$1% of polar components (or $>$50%)
- ▲ if solute still too little retained use dry hexane, perhaps on dried (activated) silica

or
- ● try alcohol—water on an ODS—silica reverse phase column, or organic (ion-exchanging) resin
- ▲ vary alcohol—water ratio

if more soluble in more polar organics like chloroform or ether —
- ▲ substitute a similar liquid for some or all alkane

if water soluble —
- ● try water—alcohol, or water—acetonitrile mixtures on silica, or organic resin
- ▲ perhaps add acids, bases or salts and vary concentration

ionic	● use ion-exchanging material (resin or bonded-to-silica) ▲ vary water, alcohol, pH and salt concentration ● or form ion-pair or other derivative
acidic or basic	● add acid for acidics, bases for basic compounds

8.2 Solvent Selection

Theory and Practice of Solvent Selection

summary
- **theory approximate guide to solvent selection**
- **solvent strength main factor controlling retention**
- **hydrogen bonding and dipole interaction affect selectivity**

Apart from its practical importance, the theory of solvent–solute–surface interactions has been the subject of legitimate interest. Long term aims of theory must be to predict suitable solvents from data on the physical properties of solvent and solute. As with GC these studies have not yet been declared practical by their popularity, but have nevertheless introduced some order, both practical and conceptual, into an otherwise rather chaotic situation. At present the rules of thumb tell the user what sort of mixtures to try first, and this is now more valuable than a definitive prediction, the ultimate aim of theory. Accurate prediction requires much more accurate data than that readily available today.

Most of the fundamental studies have come from the work of Snyder[8]. More recent reviews of the role of solvent in LSC[125], and solvent properties[126], and a practical guide to solvent selection in LC[3], update his earlier work. Rohrschneider[135] and Huber[136] have made complementary contributions. Solvent strength or polarity is linked with solubility parameters which are of much wider interest and application. Barton[128] has reviewed solubility parameters in general. Four contributions are identified and some representative classes of solute whose retention is particularly affected are:

solvent property	**examples of solvent**	**solutes which interact strongly**
high dipole moment	CH_2Cl_2	carbonyl compounds, nitriles, sulphonates, amides
proton acceptor (basic)	amines, ammonia	alcohols, acids, phenols
proton donor (acidic)	alcohols chloroform	amines, sulphonamides
polarizability (often linked to high refractive index)	tetrachloro-ethylene	aromatics, halogen and sulphur containing compounds

Solvent selectivity

Source: L R Snyder[126]

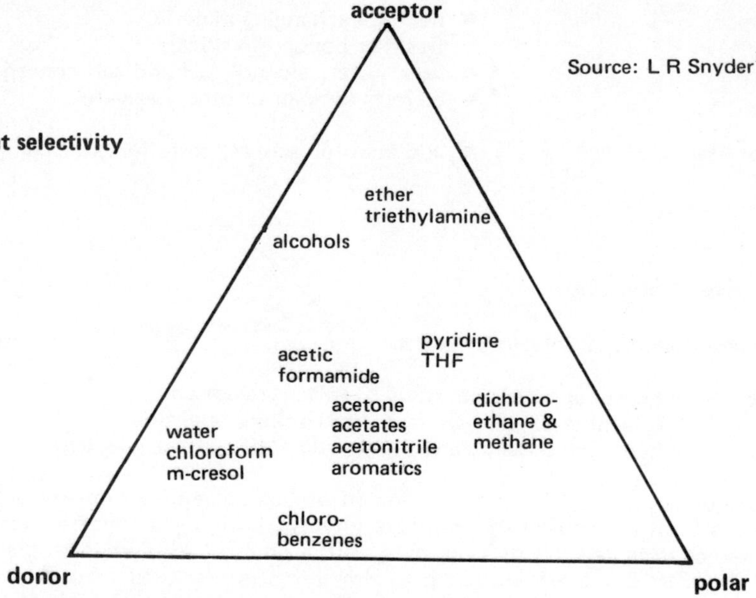

Saunders[127] has summarised briefly a few practical conclusions from the theoretical work. These help quick estimation of an approximate solvent mixture (or mixtures) to retain a compound with $k' \approx 3$, a sensible starting point.

He observes of solutes:

- saturated hydrocarbons and simple olefins are eluted with $k' < 3$ for all solutes.
- alkyl and halogen substituents change the solvent strength ϵ° needed by $< \pm 0.05$ compared to the parent compound.
- changing the number of aliphatic carbon atoms has a similarly small effect.
- difunctional groups of similar polarity are roughly additive but stronger groups mask the effect of weaker ones.
- it is difficult to predict the effect of further groups.
- polynuclear aromatic compounds should be considered as bezene compounds substituted with aromatic rings.

And of solvents:

- increasing solvent strength ϵ° by 0.05 will decrease k' by a factor of 2–4.
- decreasing ϵ° by 0.05 will increase k' by a factor of 2–4.
- selectivity greater if stronger component <5% or >75%.
- hydrogen bonding of solvent affects selectivity.

When using silica or alumina columns Snyder has proposed that only six solvents are needed to allow one to cover the complete range of solvent strength from alkanes to water. His suggestions are:

pentane (hexane), 2-chloropropane or diisopropyl ether, ethyl ether (tetrahydro furan), acetonitrile, ethanol (methanol), and water,

(with some of the author's possible alternatives in brackets). These solvents impose few limits on UV detection down to 230 nm. Other combinations are also feasible. The main problem is a gap between a non-polar solvent like pentane or hexane, and the first slightly polar one because the chromatographically suitable solvents suitable for this region also have rather poor UV transparency at short wavelengths. Perhaps it is possible to jump directly from hexane (or cyclohexane) directly to ethyl ether. Certainly a much larger number of solvents is not necessary to cover the whole range and even fewer may be suitable for a limited range.

In practice the need for a series of this type is relatively rare and many problems covering this wide a range are best tackled by a non-polar bonded phase like ODS. Commercial apparatus is not generally designed with even a six-component mixture in mind. Most real problems require more detailed separation in some regions, and selective resolution of a few components. Multicolumn systems can probably solve these problems more effectively. The key advantage of a full solvent sweep is that all components will be eluted in a reasonable time and no component will be missed because it is still stuck on the column.

Solvents for Gradient Elution

summary • **reversed phase systems with two-component gradient simple for wide range of solvent polarity**

With bonded phases like octadecylsilane it is usually satisfactory to vary the eluent composition between water and either methanol or acetonitrile. (The latter seems to be becoming more popular as it has become readily available at high purity and more reasonable price.) The pH and/or salt concentration of the aqueous component may also be altered.

The robustness of these systems chromatographically, coupled with the good matching of low absorptivity of these eluents (even reagent grades) at 254 nm, accounts for its popularity with users, especially in applications laboratories of instruments manufacturers. Other separation conditions may exist which do not require expensive gradient equipment but these conditions may also be more difficult to find and be much more specific to a particular analytical task. It may therefore only be economic to develop specific methods for routine problems.

summary
- **only six solvents needed for smooth solvent strength sweep**
- **all can be transparent to 230 nm for UV detection**

With adsorption chromatography there are more pitfalls with gradient elution: this is probably why there are far fewer published applications compared to reverse-phase gradient separations.

Snyder[125] argues that a general purpose range of solvents for gradient elution should give:

- a steady change in solvent strength
- no demixing problems—which implies not too sharp changes in strength between any two components in the series of solvents
- no sudden changes in secondary solvent effects (mainly proton donor—acceptor changes)
- low viscosity and good UV transmittance

Pairs of solvents give a smooth sweep of solvent strength over a small range but more rapid changes in strength, and perhaps changes in acidity or basicity, are also widely used and more effective for specific separations. Solvent demixing can be avoided by starting with a small concentration of highly polar additive and bringing the column into equilibrium with this eluent, and then increasing the proportion, usually approximately exponentially.

summary
- **demixing if change between solvents too great**
- **sudden changes in acidity or basicity can be confusing**

Demixing means that one component is being removed from the eluent by the packing until it reaches equilibrium. Meanwhile the column is in thorough dis-equilibrium and it is difficult to predict what will happen to compounds passing down the column. Although this effect has been exploited for excellent separations in TLC it is difficult to extend this to LC for it is essentially an equilibrium technique, and poor separation usually results. However demixing need not spoil a separation provided compounds do not emerge or move significantly during the non-equilibrium period: thus compounds of very different polarity can be separated by a step change to polar eluent after the non-polar compound(s) has eluted.

Demixing also produces more disturbance in photometric detectors than smoother gradients because sharp fronts of different composition produce rapid changes in refractive index and absorptivity. Even if absorptivities are carefully matched, the change in refractive index will produce a differential peak.

Engelhardt and Elgass[141] discuss some of the practical problems of using gradient and stress the need to run gradients without sample (but with the solvent used to dissolve the sample) to identify ghost peaks, and to identify the effect and effectiveness of regeneration.

8.3 Preparative LC

summary
- **considerable untapped potential, especially for solute identification**
- **no consensus on best particle size and column configuration**
- **large volume of solvent required (≈10 l/g solute)**

The objective of preparative LC is collection of separated material rather than measurements of quantity and retention that can be made from the chromatogram alone. By this definition, collection of the effluent from a small analytical column is just as much preparative as the use of classical column chromatography. The chromatographic difference is that resolution and separation speed may have been compromised in favour of sample capacity. The column will normally be operated under overload conditions—and the niceness of the chromatogram compromised.

Whatever separation tactic is chosen it definitely pays to develop the method on analytical-sized columns operated under preparative conditions. This saves time, sample and solvent. These columns, based on 1/4 or 3/8 in tubing may well produce enough material to answer needs for identification with NMR or mass spectrometry. If more material is required a scale-up of diameter and length with a similar packing material is usually straightforward.

Special hardware needs for preparative chromatography

summary
- **should be cheaper and simpler than analytical equipment**

Compared to analytical LC, the hardware for preparative work should be easier and simpler to make. It should also be *cheaper*: instrumentation whose price is proportional to sample weight is not cost effective. Extremely cheap, and effective, equipment is already available designed for classical column chromatography. The performance of this type of column system can be speeded up by adding a pump: this speed-up will generate a need for a detector to assess what separation has been achieved.

Pumps

summary
- **pneumatic amplifier best, >1 ml/s needed**

A flowrate capability of >1 ml/s is desirable. For this purpose pneumatic-amplifier types are most suitable. Mini versions with a stroke volume of a ml or two are excellent and cheap. Flow can be stopped with a simple tap and no over-pressure protection devices are needed to cope with column or pipework blockage. Electric motors for these flowrates tend to be large and stopping flow is more difficult. Direct gas-drive is unsafe with glass columns and only well suited to specially designed column systems such as that described by Godbille and Devaux[158,241].

Injection

summary
- **stopped-flow injection convenient**

Syringe injection with a needle is less suitable than with analytical columns because it is too easy to upset the packing at the top of the column. Stopped-flow injection is usual because it is difficult to inject large volumes from a large diameter syringe against pressure. (Typical pressures just feasible are ≈10 bar with a Hamilton 10 ml gas-tight syringe, ≈50 bar with a 1 ml size.) Nonetheless a syringe is invaluable for injecting into a suitable injection system port[238,241]. The arrangement below allows flushing the syringe to get rid of old sample, but the connection from the valve to the column must have a very small volume or it will have a serious memory of the previous sample.

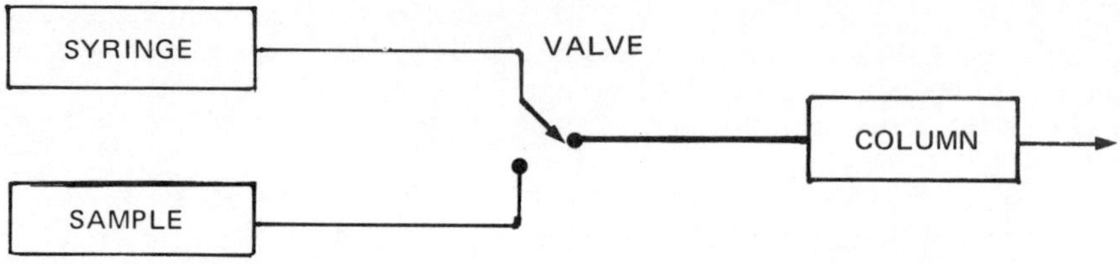

Sample injection valves (perhaps with sample loops) are most suitable, but more expensive. Scott and Kucera[138] have shown the importance of using a valve correctly to get a neat injection slug.

For repetitive injections of the same sample, some type of pump is probably most suitable. This might be the main solvent pump but a separate pump is probably more convenient. A direct gas-driven pump has been used on an automated chromatograph[238]: small-piston reciprocating pumps would also be suitable.

Distribution of injected sample

summary • **injection distribution influences effective resolution**

As with analytical LC it is not only the volume (and mass) of sample that is injected that is important, but also its distribution over the column. If we consider a 1 inch column, bore 23 mm, then a 1 ml injection should ideally be spread over the column evenly in a disc only a few mm thick. (One might also try to avoid the walls by a mm or so because wall effects cause more peak broadening there. The amount of the cross-sectional area near the walls decreases as the diameter increases, so the volume of sample that will be affected is fairly small, but tailing or peak asymmetry caused by wall effects can still have a disproportionate effect on the purity of collected samples.)

If the sample is injected in a large volume of solvent and is well adsorbed then it will tend to distribute itself evenly over the top of the column: hence this procedure is often recommended. Solubility limitations do not always allow it.

The sample distribution problem becomes more and more serious as the column diameter increases, and may well be the reason why worse results are reported for columns above 1 inch. However these problems should not be insuperable with careful engineering.

Columns

Glass columns are fairly cheap and usable up to ~10 bar. Because of the difficulty of making connections and seals, most commercial types have plungers with sintered PTFE faces which are pressed into the actual column tube. This also performs the essential function of holding the adsorbent bed so that it does not shift when pressure is applied and relieved, and thus spoil the packing structure. Glass tube can also be used in stainless steel compression fittings with ptfe ferrules but is fragile and the fittings tend to slide off, especially with pulsating pumps.

Stainless steel columns with compression fittings are most robust and reliable but modified column ends are needed at both ends to avoid dead volumes and contain sintered or woven discs to retain the packing. Engineering becomes both heavy and expensive for diameters above about 2 in if a pressure capability >50 bar is required.

Detectors

summary • **variable-wavelength UV or RI suitable**

Most UV detectors currently used for analytical work are also quite suitable for preparative LC. 254 nm types are less convenient than variable-wavelength monitors because changing the wavelength is often an effective means of changing the sensitivity. Changing the pathlength is also useful and some manufacturers provide alternative cells that can be fitted to reduce both pathlength and impedance to flow which may be rather high at flowrates of ~1 ml/s. A flowcell in fused quartz with low volume connections (Cheminert flanges screwed directly against the quartz block) with 1 mm pathlength and width and length 7 mm has been used in several standard spectrometers.

Photometric detectors can of course be most misleading because the absorptivity of compounds can vary by several orders of magnitude. So large peaks may only represent minute weights—and large quantities be almost undetected.

In practice this is a small disadvantage if spectra and absorptivity of the compounds concerned are known. It is easy to obtain spectra of all compounds with a scanning UV detector. This information is very useful when deciding the most suitable wavelength to detect at, and also provides some information to aid identification.

RI detectors are much better suited to preparative work than analytical because a much lower sensitivity is acceptable.

Flow-splitting is a complication which a suitable flowcell should make unnecessary.

Collection systems

summary • **conventional fraction collectors ideal for exploratory work**

Conventional fraction collectors are suitable. Siphons do not work well at high flowrates because they will not empty fast enough and therefore remix the separated components. The most suitable aliquot to collect varies widely with the column size, and efficiency, and the degree that a peak is retained. Thus the peak volume is about a tenth for unretained peaks as it is for components that have $k' \simeq 10$.

With a fraction collector the decision about how to group fractions together is postponed until the chromatogram is complete. This is ideal while a separation is under development: if repeated injections of the same sample prove necessary in order to purify enough sample, it will be better to decide the various collection times and cutting points.

A sophisticated computer-controlled collection control system has been described[238]. This decides the frequency with which to advance a fraction collector depending on flowrate, column size and efficiency, and time within the chromatogram, or it will automatically re-inject samples and collect up to 9 fractions at programmed positions in the chromatogram.

The collection control algorithm can be much simpler, in some ways, than that for integration of the area under peaks:

- low dynamic range
- delays in decision to collect lead to small losses which would not be acceptable for area measurement
- baseline drift should be very small
- small range of retention (k'), or the peaks will be too dilute, and therefore small variation in peak width.

But the algorithm should deal with:

- noise and spikes
- baseline noise and drift level checks
- off-scale peaks
- approximate peak heights and areas
- controlled valley cutting points
- incomplete resolution and heart-cutting.

Packing preparative columns

In principle the same methods for analytical columns are suitable for larger diameter columns. Particle size, and size distribution, density and personal preference are factors which may determine whether dry or slurry packing is used (see section 4).

In practice packing larger columns is often easier than small analytical columns, especially with slurry methods, because the volumes and diameters are more manageable. A recipe is given in section 4.2 for slurry packing large-diameter columns.

Some good features of preparative LC
(particularly compared to preparative GC)

- range of suitable compounds very high, and solubility rather than volatility needed

- sample degradation generally lower in solution at ambient temperature
- solute recovery easy by solvent removal in a rotary evaporator, or freeze-drying
- ambient temperature operation convenient and safe
- small geometry, natural to LC because of low diffusivity of liquid, easier to engineer
- convenient detectors are non-destructive and do not need splitting systems
- stop-start flow control does not reduce resolution

Some problem areas

- The optimum **particle size** (and size range or distribution) for preparative LC is not yet known, but the best ultimate compromise between a large number of factors, including production costs, is probably for spherical particles in the 10-20 μm range. Large irregular material (20-50 μm) are now cheaper, but less efficient.

- **Porous** material with a high surface loading capacity (which demands a surface area $>100\ m^2/g$—see section 6.1) and a pore size of 5-10 nm. Pellicular materials are not suitable. Bonded-phases or LLC systems may prove to have useful loading capacity but the area is as yet largely unexplored. Carbon may well prove to be very useful if a suitable material can be produced at low enough cost.

- **Solubility** in a suitable eluent is a practical limiting factor. Solvents in which the solute is more soluble are usually those that cause insufficient retention for resolution. Reverse-phase systems may offer a solution for those compounds for which there is a solubility problem with polar adsorbents[283]. Snyder's solvent parameter[126] can be used to predict solvent systems with the same eluting power ϵ° but higher solvent power.

- **Equilibration** of the solvent system by passing many column volumes of solvent is hardly feasible as it is for small analytical columns. This also places some limitations on the use of solvent changes, either stepwise or with a programmed gradient.

- The most straightforward **solvent for the sample** is the eluent. A stronger solvent may dissolve more and so permit a more concentrated solution to be injected, but upsets the column equilibration and resolution, often in a disastrous way if the sample precipitates out: a very long tail is symptomatic. The slug of strong solvent can also cause all components to be eluted quickly, without much resolution.

- **Solvent purity** is critically important. The concentration of collected samples will rarely exceed 0.1% w/v so solvent residues on evaporation should be <10 ppm to minimise contamination. This implies rather pure glass-distilled nitrogen-sparged solvents. Of course a final recrystallisation may be feasible but transfer losses mean this should be avoided if possible.

- **Peak shapes** tend to deteriorate in preparative LC. This can be caused by volume or mass overload. Tailing often has a particularly deleterious influence on the effective resolution: formulae for calculating resolution which assume peaks with a gaussian form become less appropriate. The worst case is trying to collect an impurity on the tail of a major component.

• (Injection) **volume overload** causes peaks to broaden by extending the rear of the peak[138] while the peak fronts stay in about the same place. Ultimately there is a transition from elution development to frontal analysis. Injection volume is discussed in more detail in section 8.7. A quarter of the peak volume is a useful rule-of-thumb value.

• **Mass overload** is a much more complicated and less well understood condition. Studies still in progress by Knox and Done[261], and Gray[268] have shown that column efficiency starts to decrease even at extremely low mass loading. The point at which this is apparent is only limited by the efficiency with which the column itself can be packed. With less well packed columns, overload is only apparent at much higher concentrations of sample, and masked at low sample concentrations. Apart from this effect, at much higher loads the adsorption isotherm becomes non-linear and the equilibrium between eluent and surface is disturbed, as if there is a temporary change in eluent. Peaks shapes become distorted and the front of the peak moves forward but the rear stays or moves forward much less[138]. Snyder's definition of overload[8] is arbitrarily based on a 10% reduction of retention (k').

• **Loading capacity** of a material has been based on the reduction in retention as noted above, but a better definition may be based on the rate at which efficiency is lost as a highly efficient column is loaded, at least for highly efficient columns such as those used for analytical work.

• **Fire hazards** from LC are discussed in section 12.9. Some features that reduce the danger are: pneumatically operated valves and pumps, trays to contain solvent spillages, and valves that spring to a closed position in the event of electricity or air supply failure.

• **Decomposition** of solute being purified on the silica surface, or in the solvent has sometimes been noted. Freeze-drying is recommended[260] as a very gentle procedure for recovering samples from solvents.

• **Solvent purification,** and filtration, with a **pre-column** is recommended: a surprising amount of impurities can be collected. Filtration, and perhaps pre-separation, of the sample to be separated is also desirable to avoid contaminating the column, especially for expensive high-efficiency packing materials.

• Eluents can be **re-cycled** for re-use when no peaks are detected, but impurity levels can build up and contaminate the collected samples.

• **Flowrate changes** are a convenient way of eluting strongly retained compounds if sufficient pressure is available.

• The optimum **column length/diameter ratio** has yet to be predicted by theoreticians. (In practice only a few sizes of compression fittings are available.)

• Optimal **eluent velocities** to maximise throughput have not yet been established clearly despite some work by Bombaugh and Almquist[257]. It seems likely that a reduced velocity higher than that producing optimum efficiency under analytical conditions will be feasible because losses from this cause may be masked by far greater losses caused by mass and volume overload. And of course it may be advantageous to trade resolution for throughput per time just as one compromises for analysis time under analytical conditions.

• **Permeation chromatography** may provide a higher loading capacity than adsorption chromatography because all the pore volume is used, and not just the pore surface. So there are often smaller resolution losses from mass overload[259,173]. Loads of ~1 mg/g of packing material seem about optimal. A combination of mechanism has been exploited by Pickart and Thaler[258] to carry out some interesting separations of proteins and peptides. Further exploitation of these combination separation processes can be expected in preparative LC as suitable materials become available.

• **Changing the solvent** is at least as attractive as for analytical separation problems (except that many pumping systems which offer gradient cannot pump fast enough for preparative work). The time and solvent required to re-equilibrate the column are disadvantages against which the separation gains must be set. Step changes seem likely to prove popular because some of the disadvantages like baseline disturbance are no problem if the aim is preparation and because it will be very much cheaper and simpler to engineer than a full gradient system.

Separation tactics

Currently a few tactics are in use:

• **large particles (30-100 μm), modest resolution, slow speeds**

Low resolution work uses traditional large, irregular silicas which are quite cheap, and also available in pre-packed columns. These systems speed up (and improve) traditional column methods but rarely provide plate numbers >1000 because the particle diameters are ~10 times the analytical material. For the same reduced velocity, the linear velocity can only be a tenth as fast; and for the same reduced length, the actual length must be ten times longer, so separation is much slower than on analytical columns. Only low pressures (~10 bar) are necessary to generate these flowrates. However since the weight of silica is usually quite large, the possible yield from a several hour separation run may be quite large. Kirkland and De Stefano[260] recommend this approach with a target retention of $k' \simeq 5$ for compounds of interest in their review of prep LC methods.

• **large column diameter with disposable TLC silica (5-40 μm)**
• **slow but large weight per pass possible**

Similar resolution can be obtained on TLC type silica which is cheap enough to use once and throw away: this approach is advocated by Godbille and Devaux[158]. The weight of silica is now large (~2 kg) and the volumes of solvent needed for elution are very large so it may be better not to try to elute the bands very far (keeping $k' < 5$) or even extrude the bed and cut out and extract the sections containing the compounds of interest. (One must of course be able to detect where the compounds are.) These two tactics, reducing k' and/or not eluting, reduce the time and solvent needed, but the separation speed is still low, at least with the instrumentation designed by these workers, because the pressure capability is <20 bar. TLC is recommended to guide solvent selection: the speed is too low and the solvent consumption too high to allow much experimentation on the column.

- **analytical materials (5-20 μm) give very high resolution, at high speed—and high cost**

Direct scale-up of analytical columns gives the highest separation speed but requires rather expensive packing materials and high pressures. Very high resolution is possible[241] but only at analytical loads. Because this is high speed liquid chromatography, many injections can be made in the time for one run on a lower resolution column packed with large particles: this may compensate for the low mass per injection and give a high throughput in g/hour.

An important advantage is the direct scale-up from analytical columns, and the ease of slurry packing columns. The cost of packings could be drastically reduced if there were sufficient demand. (At the time of writing there are commercial pressures against this development because it would undermine the prices of analytical material.)

In the future we can expect these different approaches to merge as instrumentation and packing materials evolve.

8.4 Quantitation

summary
- **well established methods used in GC are generally suitable**
- **concentration-sensitive (photometric) detectors need slightly different treatment from mass-sensitive detectors like flame ionisation**
- **electronic integrating systems most convenient**
- **but peak height measurement is quick and easy**

The methods of estimating actual compositions from chromatograms, destined to become only available as an electronic representation in a microcomputer, are very well known from their application to GC[17] and will therefore only be briefly summarised here.

In practice the choice is between manual peak height measurement or electronic integration of peak area. Cost, convenience and the precision required generally determine which method is used.

Manual metnods

summary
- **only peak height measurement is still common**

Comparisons show that triangulation (only reliably done by multiplying height by width-at-half-height $\simeq$ 0.94 x area), planimetry, cutting-and-weighing (or copying-cutting-and-weighing) give similar area precision[378]. (Since peaks are almost always compared with other peaks, the *accuracy* of area measurement is only rarely of configuration under measured conditions. Consider for example a 5 mm bore effects and the distribution of the injected slug by the injecting device, syringe or as the peak width precisely and this means using a reasonably fast chart speed: Ball, Harris and Habgood[378] recommend the width-at-half-height should not be less than 1/10 the height, and ideally about 1/4: this implies using more chart than normal especially if the compounds of interest have a wide range of retention (peaks are about ten times as wide at $k' = 10$ as at $k' = 1$).

All these methods are rather laborious and much less precise than electronic area measurement with an integrator, so only peak height measurement is commonly used as a routine procedure.

Peak height measurement

summary
- **simple and quick — rulers are cheap and reliable**
- **insensitive to flowrate changes for concentration-sensitive detectors**
- **sensitive to changes in peak shape, usually caused by the injection volume and distribution by injector**
- **sensitive to changes in retention, or tailing**

The main merit of peak height measurement is that it is quick and easy, and does not require an inconvenient amount of chart paper. For concentration-sensitive detectors like UV, it has the added advantage that changes in flowrate cause quite small errors, whereas area measurement is quite sensitive to deficiencies in the pumping system.

On the other hand, peak heights are affected by the injection distribution, usually the main cause of changes in peak shape: this is particularly a problem with on-column injection giving very high efficiency on short columns packed with micro-particles. A consistent peak shape is essential and usually this also implies a fairly consistent injection volume.

Electronic integration

summary
- **precision of integrator is better than chromatographs**
- **convenient but expensive**
- **cost most easily justified by extra data processing facilities, for example to produce a printed report**

Electronic integrators, all computers programmed in hardware or software with algorithms of varying degrees of complexity, now provide a convenient method of measuring both area and height with a precision and accuracy greater than that of the chromatograph itself. The pump, injection system and detector are now the main sources of error. (See also section 7.7.)

Calibration

summary
- **calibration nearly always needed**
- **finding a pure sample for calibration is often difficult**

Because the response factor of most detectors can vary very widely (even the transport system—FID detectors do not give factors as near unity as the FID in GC), calibration is almost always needed. Only rarely can response factors be assumed equal, particularly if the detection wavelength is not chosen to try to equalise sensitivity. Many compounds tend to have more similar specific absorptivity at short (<230 nm) wavelength. Derivatization has been used as a way of equalising response factors: the specific molar absorbance of NBCl—digitalis glycosides is effectively a function of the number of derivatised hydroxyl groups in the molecule[278].

The need to calibrate is a serious practical problem when dealing with unknown compounds, or ones which are unstable or difficult to purify. A knowledge of the structure or at least the compound type together with a UV spectrum may allow an approximate estimate of the specific absorptivity, perhaps accurate enough to assess the level of a trace impurity. Otherwise a pure sample is needed for an accurate answer. Preparative LC will often be a good method for preparing a small pure sample, perhaps from a sample which has a high concentration of impurity by selecting a suitable impure sample or by synthesis, or accelerated decomposition or degradation (see section 8.3).

Errors from changing flowrate and peak shape

summary
- **flowrate changes peak area (with concentration-sensitive detectors)**
- **retention and efficiency, injection distribution and tailing changes peak height**

Intuitively the effect of flowrate changes on peak area and height with concentration-sensitive detectors should be clear from the following example. (A detailed treatment has also been given[380].)

If a peak is passing through the detector and the flow stops, for example while the pump recycles, the peak height will be unaltered but the area will continue to increase even though no liquid is passing and the chromatogram is not developing further.

On the other hand if a peak is taller because it tails less, or is eluted sooner, the area measurement will compensate but not the peak height.

Quantitation with Gradient elution

summary • **requires very accurate pumping system**
• **avoid if possible**

The error analysis of systems using gradient elution is a little more complicated but has been discussed, with experimental verification, by Bakalyar and Henry[380]. The fundamental need is still to control the total flow rate when area is measured, and the composition (and thus retention) for constant peak heights.

It is therefore necessary to control the flowrate of *both* pumps precisely to control both total flowrate and composition, and not only during one chromatogram but consistently for all chromatograms. (The requirement for absolute flow consistency is more stringent than that for isocratic operation when ratios of peak heights or areas are affected little by changes in flowrate *between* chromatograms, for example as column frits clog up slightly.) This makes considerable demands on the pumping system and suggests avoiding gradient elution whenever possible. Very roughly a 1% flowrate error by either pump is likely to cause an error of around 1/2% in composition.

Standards

summary • **internal standards needed for syringe injection**
• **duplicate valve injections may allow external standardisation**

As with GC, internal standards are widely used to compensate for a number of variables, most critically the injection volume. Internal standards (that is compounds added in known concentration to the solution to be injected) are almost essential for precise and credible results with syringe injection. The compound(s) chosen for internal standard(s) should ideally have a retention intermediate or bracketing the peak(s) to be measured, and should of course be readily soluble in the eluent, stable, non-volatile and pure, and easily weighed out.

Some injection valves may be precise and reliable enough to rely on the absolute injected volume. Duplicate injection which replicate and periodic injection of (external) standard solutions are still needed to produce trustworthy data.

8.5 Trace Analysis

summary • **LC excellent for trace analysis of non-volatile compounds**
• **the need to separate reduces ability to detect trace components**
• **peak height measurement is usual for trace analysis**

Trace analysis techniques using LC have been discussed in some detail with many examples by Kirkland[351], and in his book with Snyder[3]. Karger, Martin and Guiochon[199] have given a detailed mathematical treatment of the problem of maximising sensitivity to trace components, with some experimental work demonstrating the importance of using large injected volumes. They sum up the nub of the problem thus:

"Chromatography is included in an analysis scheme when it is necessary to separate the components prior to detection . . . The best detection limit is obtained when no column is used, but, of course, all separation is then lost."

The column then is a means of separation, but also a way of diluting the sample making it more difficult to detect. The practical problem is how to obtain sufficient separation with the minimum of dilution. Theory helps selection of the most practical solution.

Peak height measurement is nearly always used for trace analysis because it is less sensitive to variations in resolution from adjacent, perhaps overlapping, bands. The precision of a measurement of a trace component is not normally expected to be much better than 5%. (Area measurement by contrast gives a higher precision because it uses more of the data, and is usual when a more precise measurement is required.)

The importance of the 'quality' of the detector is less than one instinctively imagines, apart from the obvious points like avoiding RI detectors. The noise of the most expensive is only several times less, and certainly not an order of magnitude less than the cheapest. In contrast, careful choice of column and system and operating method may give an improvement of two orders of magnitude. Noise free pumps also make a small contribution to improving detection limits; gas driven coil pumps are especially quiet.

summary • **column dilutes injected samples**
• **dilution must be minimized to improve detection limit**

The concentration at the peak maximum c_{max} is given by

$$c_{max} = \frac{m\sqrt{N}}{V_R\sqrt{2\pi}}$$

where m = mass injected
N = plate number
V_R = retention volume of component

(The factor $\sqrt{2\pi} \simeq 2.5$ assumes a Gaussian profile, and is a sufficiently precise estimate for this type of calculation.)

If for the mass m, we substitute $c_o V_o$, where c_o is the initial concentration and V_o the injected volume, then

$$c_{max}/c_o = \frac{V_o \sqrt{N}}{V_R \sqrt{2\pi}}$$

c_{max}/c_o represents the amount that the injected sample is diluted. If equal to 1, there is no dilution; and if zero, it has been infinitely diluted. For the highest sensitivity we need to maximise c_{max}/c_o.

This simple formula can be used to estimate the dilution of a particular column configuration under measured conditions. Consider, for example, a 5 mm bore column, 100 mm long packed with 5 μm particles. The retention volume and efficiency of a component can be measured from a test injection. If the retention volume is ~3 ml and the column has a plate number of 5000 for an injected volume of 10 μl, then

$$c_{max}/c_o = \frac{0.01 \times \sqrt{5000}}{3 \times \sqrt{2\pi}} \simeq 0.1$$

So the sample is diluted tenfold by the column. Knowing the absorptivity of the sample at the detection wavelength, and the detector noise level, the minimum detectable concentration can be calculated easily.

Some ways of decreasing the dilution are obvious. One is to use a more efficient column, increasing the plate number: another is to increase the injection volume; or to reduce the retention volume V_R. Obviously all these variables interact. For example, if N is increased by using a longer column, then V_R will also increase; as the injected volume V_o is increased, the plate number efficiency N will eventually fall.

To take these interactions into account and thus predict the optimum geometry, a more detailed treatment is needed: this is given by Karger, Martin and Guiochon[199], and is summarised in a simplified form below.

Injection volume

summary
- **increasing injection volume is single most important method of improving detection limit**
- **increase from at least a fifth, and up to twice peak volume**

In practice, the effect of injected volume on efficiency is complicated by wall effects and the distribution of the injected slug by the injecting device, syringe or valve, and connections. As the injected volume rises the infinite diameter effect will probably be lost. Since the fraction of packing material that is near the wall is less for larger diameter columns, loss of efficiency will probably be less with these columns. On the other hand the problem of delivering a plug-shaped injection slug increases with larger bore columns. With existing injector designs, bores between 6 and 10 mm seem to give rather better results, especially because tailing is often reduced.

Without further detailed experimental work, the theoretical treatment of Guiochon and co-workers cannot be applied very precisely. Nonetheless an invaluable general rule that emerges is to start with a fifth of the peak volume and investigate increasing this up to twice the peak volume (see section 8.7)

[peak volume = flowrate x chart distance/chart speed]

A constraint on increasing volume loading may be that the column efficiency falls from *mass overload* caused by the *major* components. A reduction in retention volume of the main peak and change in shape are symptomatic.

If the injection volume is chosen so that the zone broadening due to the detector is a fixed fraction of the overall observed broadening, then the useful rules of thumb given in section 8.6 can be derived.

To inject large volumes it is usually most convenient to stop the eluent flow unless a loop injection valve is being used. Syringes (with PTFE-tipped plungers) remain most convenient for trace analysis.

Retention volume

summary • **to reduce diluting power of column**

- **use a short column efficiently packed with very small particles**
- **use large bore columns, preferably 6-10 mm**
- **choose a selective packing and eluent system**
- **adjust k′ in range 0.5-2**
- **operate so reduced velocity ν is 1-5**
- **try to elute small peak first**

The retention volume can be calculated from the equation

$$V_R = \epsilon \pi d_c^2 . L. (1 + k')$$

To minimize V_R we should use as short a column as possible (consistent with keeping the same plate number) and the peak should be as little retained as possible, keeping k' small (consistent with adequate separation).

For a given sample size, the column cross-sectional area should be as small as possible for a given injected volume. In practice a larger diameter can be exactly compensated by a larger injected volume without any loss of efficiency. The diameter should be chosen so that the peak volumes are large enough compared to the detector cell volume to avoid loss of resolution and more important here, loss of peak maximum concentration (see sections 7.6 and 12.4). For very short columns <100 mm, bores rather larger than usual for quantitative analysis, say ~8 mm bore will give slightly better results.

Only if the actual mass (and volume) of sample is limited should very narrow columns (<5 mm) be used, and in that case, very small detector cells (say 2 μl volume) are desirable.

The porosity of columns is sensibly constant in practice. (Although pellicular materials have a much smaller porosity, their low efficiency rules them out for trace analysis.)

Since the highest efficiency is needed (giving the largest plate number N), the column should be operated at its most efficient (reduced) velocity, say such that $\nu \simeq 5$ (see section 3 where suitable flowrates are tabulated). If extremely small particles are used in very short columns, for example 3 μm particles in a 50 mm column, then this requirement may conflict with giving enough time for the detector to respond and some compromise must be struck.

Reducing the length L without losing plate number N is only possible by using smaller particles (assuming all columns are 'well packed' with reduced plate heights ~2 under test conditions). To keep the plate number constant, the reduced length l must also stay constant. Since $l = L/d_p$, L can only be reduced if the particle diameter d_p is correspondingly reduced. (An even stronger dependence of detection limit on d_p is found when the effect of resolution is taken into account[199]).

The amount a component is retained, described by the column capacity factor k′, is one factor controlling the degree of separation from other components. Because the term 1 + k′ also controls the retention volume, so

$$c_{max}/c_o \propto 1/(1 + k')$$

(If column efficiency is lower for well-retained compounds, for example with organic resins in which diffusion is slow, the dependence is even stronger.)

For any separation system, adjusting k′ by altering the eluent composition slightly controls the compromise between separating and detecting. If the eluent system is chosen so that there is more separation than necessary, sensitivity to trace components will be lost. It is usual to adjust conditions so that k′ lies between 0.5 and 2.

The other important factor determining resolution is the selectivity of the chromatographic system. It can be shown that

$$c_{max}/c_o \propto a - 1/a$$

for constant column efficiency and geometry, resolution R_s requirement and k′. When the components are difficult to separate and a is near unity, the dilution effect depends critically on the selectivity: for example, if a is improved from 1.01 to 1.05, the dilution is reduced by a fifth. Thus choosing a packing material and solvent system that gives the most *selective* separation is an important step towards sensitive detection of trace components. This one factor may over-ride other considerations about column geometry.

Another factor, ignored in the above theoretical analysis which assumes symmetrical peaks, but important in practice is that peaks usually tail. So it is easier to measure a small peak that elutes before the major component. Careful choice of chromatographic system may allow one to select the best elution order.

[Note that the above analysis ignores the variation of plate height with velocity discussed by Guiochon and co-workers. In practice, current technological limits prevent the use of very small particles in very short columns except in the zone where plate height depends only marginally on velocity. Indeed the limit on reducing particle size or length further is imposed by the speed of detector and recording, and by the engineering problems of injecting samples and transporting the eluted peaks to the detector without broadening, or tailing.]

summary
- **concentration, perhaps using the column, may be needed**
- **separation by clean-up or detector selectivity may reduce demands for separation by the column**
- **danger of artefacts in all trace analysis**

Trace analysis problems vary widely. In most cases, for example the analysis of polluted water, there is almost unlimited sample available, but at very low concentration. A rather more difficult class of problem is where the sample size is strictly limited, for example samples of blood in which the concentration, and even breakdown pathways, of an administered drug must be assessed in the presence of large amounts of many different interfering substances. There is a clear limit to the amount of 'sample' the patient can spare.

In the former case concentration is probably needed more than separation—which is the factor critical to the success of the latter type of analysis.

Concentration is of course only a special form of separation. Chromatographically it can only be achieved by the almost unchromatographic method (since the journey down the column involves dilution) of adsorbing a small component onto the top of a column under conditions where it is not eluted. This concentration can then be eluted with a stronger solvent into the detector for measurement of the amount accumulated. Many examples are tabulated by Walton[132].

A typical case could be the accumulation of non-polar impurities in water on a non-polar stationary phase like ODS-Permaphase or on carbon. Adsorbed organics could be eluted and perhaps separated with an organic solvent like methanol, perhaps using a gradient. Similar methods are already widely used in GC.

Many column switching techniques, as discussed in section 8.9, can be put to use, especially back-flushing, if, for example, there is no requirement to separate the concentrated organic components.

If separation or 'clean-up' is the primary problem, then the separating power of the system is most important, unless the problem can be circumnavigated by making the detection system more selective, and reducing the need for the column to separate. Extraction, pre-column derivatization (see section 8.10), column switching (section 8.9), and selective detectors employing variable wavelength UV, fluorescence, and radioactivity, and post-column derivatization (section 8.10) may provide solutions to these very individual problems.

All of these techniques carry a special danger of producing artefacts and it is vital to verify the identity of peaks either on-line or after collection, or at least provide evidence by 'spiking' with known samples at known concentrations.

8.6 Minimum Detectable Sample Mass and Concentration

summary
- **minimum detectable concentration is 10 times noise equivalent conc**
- **minimum detectable quantity in grams typically equal to noise equivalent concentration in g/ml**

These parameters can be determined from the detector noise c_{noise} in sample concentration terms (rather than in terms of absorbance or refractive index), and the peak volume. c_{noise} can easily be calculated for the substance concerned, taking account of the noise in absorbance terms, absorptivity at the chosen wavelength, or refractive index difference of the solvent and solute. Typically for uv detectors c_{noise} is 10^{-12} g/mm^3 or 10^{-9} g/ml (for $\epsilon = 10^4$); and 10^{-6} g/ml for RI detectors.

If a peak has standard deviation σ, and thus peak volume 4σ, and concentration c_{max} at its top, the quantity of material in the peak is

$$\sqrt{2\pi}\,\sigma\, c_{max} \simeq 2.5\,\sigma\, c_{max} \qquad (\sigma \text{ in volume units})$$

For signal/noise ratio of 4, $c_{max} = 4 \times c_{noise}$

Thus the minimum detectable quantity $= 4 \times 2.5\,\sigma\, c_{noise} = 10\,\sigma\, c_{noise}$

If the peak volume is 100 μl at $k' = 0$, then $\sigma = 25(1+k')$ mm^3

$$\text{minimum detectable quantity} = 10 \times 25(1+k')\, c_{noise}$$

If $k' = 3$, $\qquad = 1000\, c_{noise}$

For a uv detector with c_{noise} 10^{-12} g/mm^3, this quantity is 10^{-9} g, which suggests that an easily remembered rule is 'that the minimum detectable quantity in grams is roughly equal to the noise equivalent concentration in g/ml'.

The minimum detectable concentration is given by

minimum detectable quantity/maximum injectable volume.

If the maximum volume for 10% resolution loss is 1/4 the peak width, or σ

$$\text{minimum detectable concentration} = 10\,\sigma\, c_{noise}/\sigma = 10\, c_{noise}$$

which applies regardless of column geometry[103].

8.7 Injection Volume

summary
- **to keep resolution with infinite diameter columns limit injection to 1/10 peak volume**
- **for sensitivity or prep inject up to 1/4 peak volume**

Volumes that can be injected without resolution loss vary widely with column geometry and retention, and are larger than those instinctively chosen by most chromatographers. It is only necessary to know the peak volume to mentally calculate approximate injection volume limits. The peak volume can be estimated from the peak base width (4σ) in μl obtained from the flowrate and chart speed.

Since $\sigma^2_{obs} = \sigma^2_{col} + \sigma^2_{inj}$ where σ_{obs} is the observed variance

and the condition for 10% resolution loss is $\sigma_{obs} \leqslant 1.1\,\sigma_{col}$

thus $\sigma^2_{obs} \leqslant 1.2\,\sigma^2_{col}$

$\sigma^2_{inj} \leqslant 0.2\,\sigma^2_{col}$

$\sigma_{inj} \leqslant 0.46\,\sigma_{col}$

If a peak has base width 100 μl, then σ_{col} is a quarter of this, 25 μl, and the injection volume may be up to 12 μl. This simple analysis ignores the rather important distribution of the injection slug.

In practice then, allowing a margin to keep the infinite diameter effect, one may inject about a tenth of the peak volume. Where sensitivity or loading is more important, up to about one quarter of the peak volume will not degrade resolution excessively. Note that the actual volumes increase with increasing retention. If in the above example the peak volume was 100 μl at $k' = 0$, changing solvent so that $k' = 3$ would increase peak volume by a factor $1+k'$, that is fourfold to 400 μl, when over 100 μl could be injected[103].

Distance over which injection may be spread

summary
- **injected sample occupying $\sqrt{N}$ plates causes negligible loss**
- **occupying $6\sqrt{N}$ plates causes 30% resolution loss**

An equivalent and complementary way of looking at the injected sample distribution and its effect on zone spreading is to calculate the number of plates that an injected zone may occupy as a function of eluted peak broadening. This allows assessment of the *distance* down the column over which the injection may spread.

Giddings[1] has derived the following equation (no. 2.14–31)

$$\text{no of plates occupied by injected sample} = \sqrt{32 \times \text{broadening factor}} \times \sqrt{N}$$

Approximate values are tabulated below. (Note that because real slugs are never gaussian in profile, or even square, differentiation between the respective factors, 32 and 24, is not justified.)

peak broadening %	plates occupied by injected sample
5	$\sqrt{N}$
10	$2\sqrt{N}$
20	$3\sqrt{N}$
50	$4\sqrt{N}$
100	$6\sqrt{N}$

For a 5000 plate column, $\sqrt{N} = 70$ so we might reasonably choose to inject over twice this distance or 140 plates. For a well packed column of 5 μm particles with three particles per plate (reduced plate height = 3), the plate height would be 15 μm and 140 plates equivalent to 2.1 mm. From this distance the volume can be estimated taking the porosity as 3/4 for porous particles and 1/2 for pelliculars, but clearly it is simpler to derive the injection volume directly from observed peak volume and to only use this method to assess the geometrical precision with which the sample must be injected. As smaller particles are used, and the efficiency of packing methods improves, the smallness of this distance causes some difficult engineering problems which limit the size of particle that seem likely to be useful to >1 μm[140]. It is the small distance that is as much the problem as the volume.

The effect of peak volume

Column: Spherisorb S2-20W, 200 x 23 mm bore
Eluent: hexane:acetonitrile (99:1 v/v)
Injection: syringe on-column, stop-flow
Detection: UV at 250 nm

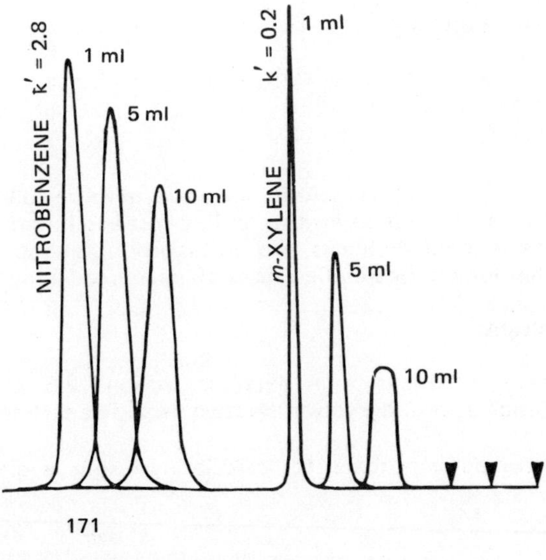

Superimposed peaks from 1, 5, and 10 ml injections containing the same weight of solutes.

8.8 Column Performance Assessment

summary • **careful testing needed to make valid comparisons possible**

It is clear that comparison of columns and their packing methods is made difficult, and sometimes invalid, by differences in column testing procedures. This section proposes some guidelines for selection of test solutes and solvents: those parameters that should be recorded; methods of calculation and presentation of results, and their interpretation.

Test mixtures

It is useful for a test mixture to have the following types of components[52]:

- An unretained (but not excluded) solute for assessment of the volume between the particles and in the pores. A compound very similar to a component of the eluent is usually best even if it only shows up as a derivative peak on a UV detector. Alkane homologs are suitable with alkane solvents; D_2O with aqueous systems.

- A minimally retained component with $k' \simeq 0.2$ to assess zone broadening caused mainly by the injector, column and detector. The peak volume of this component will be small and thus be the most critical test of these aspects of performance.

- A moderately retained component with $k' = 1 - 3$.

- A well-retained component with $k' = 7 - 20$, perhaps optionally. This component will have a large peak volume so deficiencies in the detector and injector have less effect: zone broadening is mainly a measure of the column performance.

- An excluded component, and perhaps partially excluded components.

- Solutes that test selectivity and zone broadening from thermodynamically controlled processes rather than the kinetically controlled broadening from the column packing method.

It is also useful for solutes for a test mixture to be readily available at high purity, to be stable and unreactive, of low cost and low toxicity. Simple small molecules of similar molar volumes and diffusivity (preferably known experimentally) focus attention on the kinetic causes of zone broadening.

Solvents

These are discussed in detail in section 12.5. Unless there are some over-riding considerations, the solvent system should be selected from the 'first choice' list.

A computer program for calculating results is given in section 12.10, page 257.

8.9 Column Switching

summary
- **changing column most useful for repetitive analysis**
- **little exploited for LC compared to GC**

When mixtures containing components whose retention is very different need to be analysed straightforward elution with a single column and a single eluent is often too slow and may not give effective resolution. Changing the eluent composition stepwise or with a gradient can elute a wider range but the delay for column re-equilibration and high equipment cost make this less attractive for analysing many samples. Those faced with the analysis with GC of messy real-life sample have developed a number of column switching techniques which are simple, cheap and reliable, and are now widely used in process analysis. In principle many of these systems can be extended to LC but in practice there are problems, mainly with the switching device, and so far few applications are reported.

There are far too many possible configurations to describe each in detail. Each configuration will be special to its application with the aim of speed, resolution, or grouping peaks together into classes of compounds. One can consider changing:

- column length and number
- column activity (surface area) or character (bonded phase)
- flow direction and speed
- eluent composition

In this section only systems that use a single eluent will be discussed. Dolphin et al[93] show how powerful this technique can be when combined with column switching to analyse simultaneously for fats and pesticides in milk. Suatoni and Swab[398] report successful use of backflushing to measure the quantity of all olefins, all saturated alkanes, and all aromatics in petroleum.

Flow-switching Valves

summary
- **flow control by valves often causes some peak broadening and tailing and requires specially designed valves**
- **standard valves outside the flowstream (Deans system) can be more efficient but for a limited range of configurations**

For LC, the problems of designing a valve for flow switching are different, but more difficult, than those in GC. Although ambient temperature operation is usual the pressures in use are far higher. The key problem is that the peak volumes are so much smaller so that the zone broadening caused by the valve must be much smaller than a valve suitable for GC. Huber and co-workers[227] describe a valve which, by careful design, nearly meets the needs of the very efficient columns now in use with very small particles, and a commercial version was used in the system described by Dolphin[93]. (Only the problem of designing injection valves is greater—see section 7.4.)

To be successful the loss of resolution caused by valves must be small compared to the overall gains in separation performance.

Deans[230] concluded that the use of valves to route the gas flow was a problematic solution because of the zone broadening, temperature and reliability requirements. This led him to devise systems to backflush[230] and heartcut[231] with only valves *outside* the oven. Because these valves were not in the gas flowstream carrying the solute they could not cause zone broadening, and need not be specially designed. This concept can be extended to LC with the added advantage that if flow stops in any column, peaks do not broaden perceptably by diffusion as they do in GC.

Only a small reduction in plate number is caused by the configuration shown below:

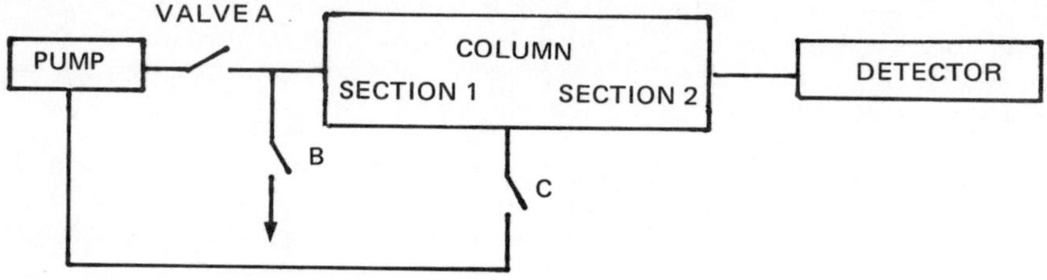

The main reason is that the two sections of the column can be slurry-packed in one go, with the valve C off. If A is open and B and C shut the column operates like one normal column. If after a set time, chosen to suit the problem, valve C is opened, flow in section 1 stops: any peaks which have not yet reached the tee at valve C stop while those which have passed are eluted normally. The peaks captured in section 1 can be backflushed out via valve B, perhaps through a detector. Or the peaks can be eluted forward after a delay to allow section 2 to separate small early peaks without complication from overload caused by major peaks.

Practical problems in implementing this system lie in the design of very low-volume end fittings which prevent the column being disturbed by reversing its flow. The tee was constructed from a compression fitting with a woven stainless steel mesh disc (section 10.7) to prevent any packing from escaping from the tee.

Extension to heart-cutting[231] is possible:

or the heartcut might be separated and monitored on a separate column

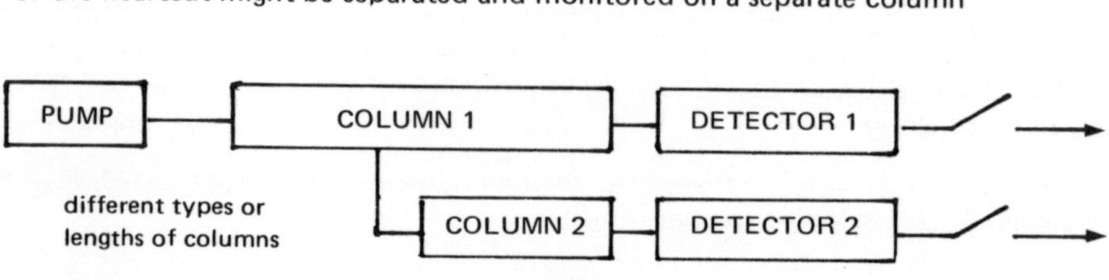

The chromatogram below shows the basic configuration in operation on a simple test mixture.

Backflushing

Conditions: eluent 1:99 acetonitrile in hexane at 1 ml/min; UV detector 254 nm 2A fsd

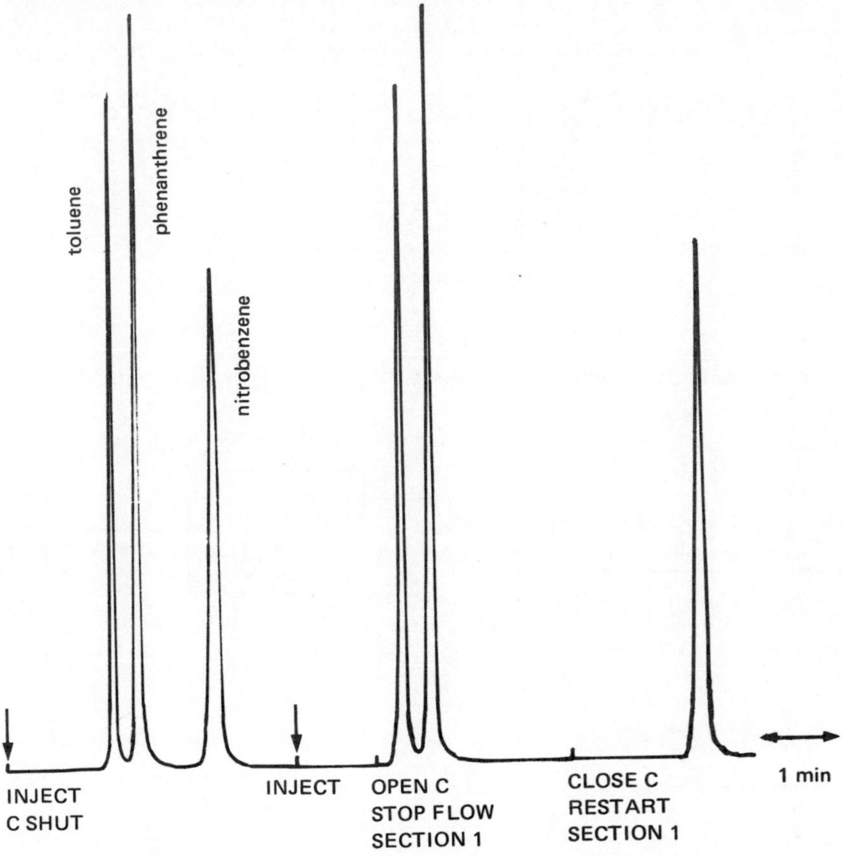

Flowrate changes

summary • **easy to engineer but limited effect on elution time**

Although changing the flowrate can rarely give more than a tenfold reduction in retention time, it is often very easy to engineer, particularly with pneumatically driven pumps. A typical implementation would be to increase the driving air pressure with a selection valve linked to two pressure regulators in order to elute stubbornly retained peaks more quickly. Disturbances to column and detector are small and very temporary, even from step changes.

Some possible configurations with flow-switching valves are:

Multiple column types or lengths

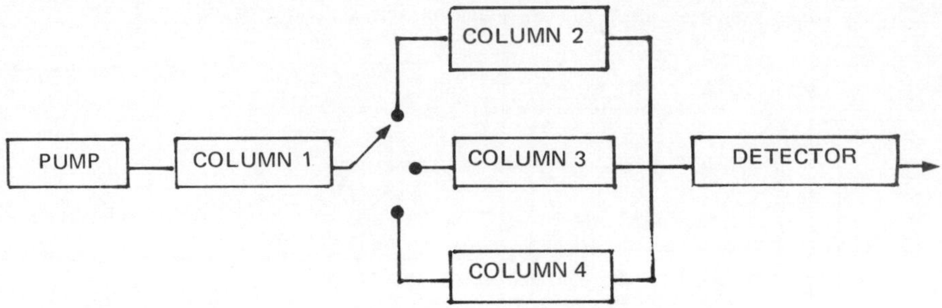

different types or lengths of columns

Separating compounds of widely differing retention without gradient elution

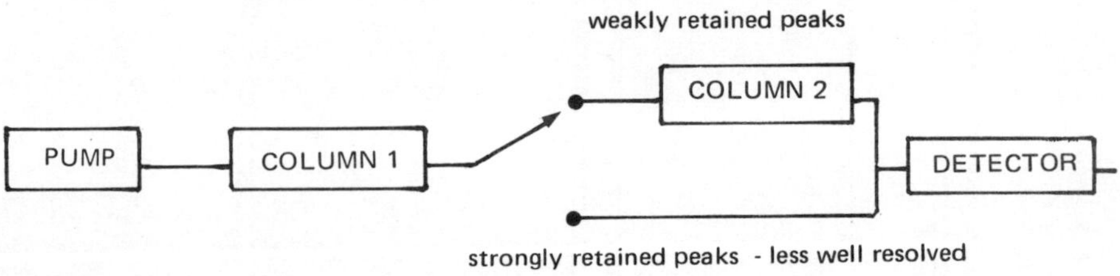

Detecting back-flushed peaks

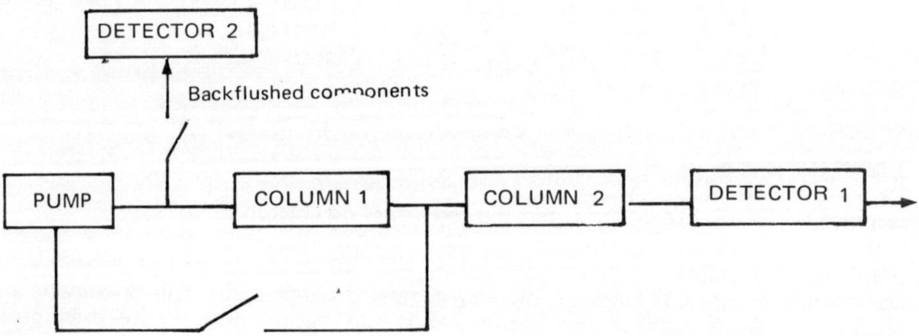

8.10 Derivatization

summary
- **improves sensitivity and selectivity of detection**
- **can change chromatographic behaviour—usually much less polar**
- **can be pre- or post- column**

Somewhat unexpectedly, derivatization is proving an essential step in solving many practical analytical problems with LC. The reasons for derivatization are different from those in GC (which is perhaps why its popularity is unexpected). Frei and co-workers have exploited and reviewed the technique[270].

The primary aim of derivatization in LC has been to allow both more *sensitive* and more *selective* detection.

The most typical example is the coupling of a fluorescent group to an amine or hydroxyl group. With the right selection of excitation and emission wavelengths, very small concentrations of fluorescent compounds can be detected, while all compounds that do not fluoresce under those conditions provide very little detector response.

Some interesting chromatographic effects are also emerging. Most derivatizing reagents link to highly polar groups, so derivatization has the effect of reducing the polarity. This may permit the use of adsorption chromatography rather than reverse-phase systems, and an easy separation from underivatized molecules. The molecular features of the derivatized molecules which allow separation are of course different from the parent molecules.

Indeed this selection may take place during the derivatization procedure in that only *less polar* derivatized molecules may be extracted for injection. This provides a valuable built-in clean-up step for analysis of messy samples like polluted water, pharmaceutical and agricultural formulations, and samples of biological origin like urine and plasma. In just these samples, the concentration of molecules of interest is usually very low. Specific extraction may also be a concentrating step.

summary
- **more sensitivity by**
  - **linking chromophore with higher absorptivity**
  - **or fluorophore to re-emit light**
  - **concentration by extraction**
- **more selectivity by**
  - **linking distinguishing group**
  - **selective extraction**
  - **polarity change helping separation**

Pre-column derivatization takes place before separation on the column: post-column derivatization is a reaction with the separated solutes from the column before it reaches the detector.

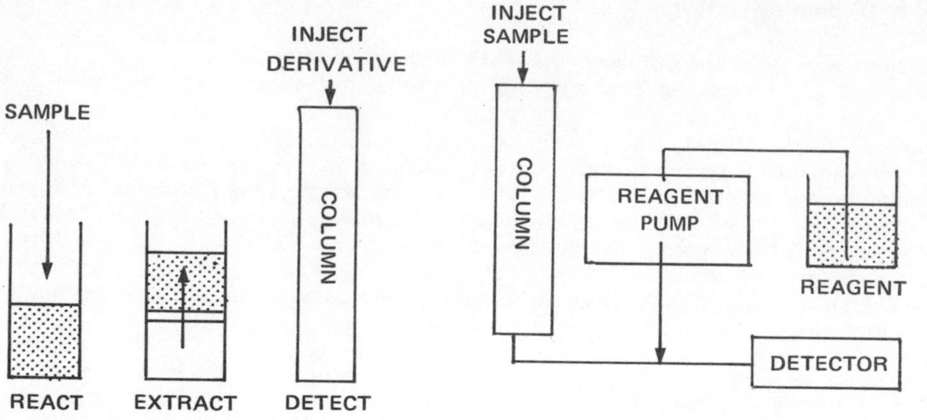

pre-column derivatization

post-column derivatization

Derivatization and Fluorometric Detection

Detector: Perkin-Elmer Model 1000 fluorimeter
with 20 ul square section quartz flowcell
Excitation: 337 nm, Emission 430 nm
Column: Spherisorb S5W, 250 x 4 mm bore
Injection: Valco loop injector, 10 ul
Internal Standard: 1-methyl, 1'-butyl, 4,4'-dipyridylium ion
Derivatization: plasma sample made alkaline, oxidised with ferricyanide, and extracted into organic solvent.

use of structurally similar compound as internal standard minimises errors due to incomplete derivatization or decomposition

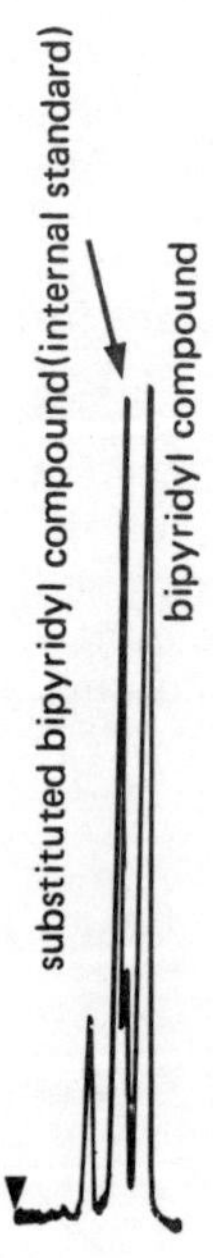

Source: P B Banham and W J D Laird, ICI Central Toxicology Laboratory, see also I Jane and J F Taylor[393] and unpublished work.

Pre-column derivatization

summary
- **can be slow and simple, and help clean-up**
- **changes chromatographic separation**

Pre-column derivatization can be a deliberate reaction step to transform the molecules for example by adding a fluorophore, or it can be a less conscious step like forming an ion-pair or free acid from a salt by injecting a sample into the eluent. Despite the differences between these two techniques, it seems consistent and convenient to consider them together in this section. The key factor is that the molecule of interest is modified in some way before being separated. In order to plan or understand the separation, it is obviously desirable to know what molecular species is being separated, regardless of whether it is formed reversibly or irreversibly.

Some advantages of pre-column derivatization are:

- can be slow, and single or multi-step
- can act as a clean-up or selective extraction step
- no restrictions are imposed by the solvent system
- excess reagents can be eliminated
- separation can be improved by changing chromatographic properties
- a wide choice of reagents is possible

but

- there is a danger of forming artefacts
- several derivatives can be formed from one compound
- incomplete, or inconsistent reaction can make quantitative assays difficult.

In common with most trace analysis methods, it is essential to run blanks and standards to guard against wrong results.

Examples of pre-column derivatives

reagent	reference	solute type	detection
danysl chloride	308	amino acids, phenols, amines, peptides	fluorimetry ex 340, em 510
phenacyl bromide	95	long-chain fatty acids, phospholipids, prostaglandins	UV photometry
benzoyl chloride	210	carbohydrates (to perbenzoylates)	UV photometry
fluorescamine	170,193,195 Fluram	amino acids, peptides, primary amines	fluorimetry ex 395, em 490
N-succinimidyl p-nitrophenylacetate (SNPA)	Regis Chromotag	amines, amino acids	photometry 254 nm
3,5-dinitrobenzoyl chloride	Regis Chromotag	alcohols, phenols, amines	photometry 254 nm
p-nitrobenzyloxyamine (hydrochloride)	Regis Chromotag	ketones and aldehydes	photometry 254 nm
1-*p*-nitrobenzyl-3-*p*-tolyltriazine	Regis Chromotag	carboxylic acids	photometry 254 nm
4-chloro-7-nitrobenzo-2,1,3-oxadiazole (NBD-Cl)	278	glycosides, aglycones (-OH to 4-nitrobenzoates)	photometry 260 nm
phenyl hydrazine	279	ketosteroids	UV photometry
dansyl hydrazine	313	carbonyl compounds	fluorimetry
N,N′-dicyclohexyl-carbodiimide (DCC)	307	fatty acids → *p*-Methoxyanilides (other reagents used to prepare same derivatives)	UV photometry

Some applications of post-column derivatization

authors	ref	solute type	reagent	detection
Higgins	321	benzoyl chloride	sapogenins	photometry 325 nm
Nakajima Ozawa Tanimura Tamura	275 309	carboxylic acids	dicyclohexylcarbodiimide (DCC) and hydroxylamine perchlorate (HAP) coupling to form hydroxamic acids; then Fe^{3+} complex with ferric perchlorate	photometry 530 nm
Morrison, Lou Hamilton	266	carbohydrates	orcinol/sulphuric acid	photometry 420 nm
Samuelson Johnson	277	dicarboxylic acids	chromic acid oxidation	photometry
Spackman Stein and Moore	167	amino acids	ninhydrin	photometry
Maeda and Tsuji	273		0.03M KOH/MeOH (permeation chromatography)	fluorimetry
Mays, Apeldoorn Lauback	276	kanamycins	*o*-phthaldialdehyde (*o*-PTH) and fluorescamine	fluorimetry ex 320, em 450 nm ex 375, em 480 nm
Roth	228	amino acids	*o*-PTH	fluorimetry ex 340, em 455 nm
Muuze and Huber	310	phenothiazines	acid $KMnO_4$ oxidation; H_2O_2 to destroy excess reagent	fluorimetry
Katz and Thacker	145	carbohydrates	H_2SO_4	photometry ~300 nm
Wolkoff and Larose	280	phenolic compounds	Ce^{4+} sulphate	fluorimetry

Ion-pairs

summary • **many very polar molecules do not chromatograph well**
• **ion-pairs are reversibly formed derivatives**

Many molecules, particularly many of biochemical interest, contain very polar groups which make them difficult to chromatograph satisfactorily.

The vast range of sulphonic acids used mainly for making dyestuffs are typical molecules of this type. The sulphonic acid group is probably able to displace even water from the surface of silica. The observed results are very broad and tailing peaks for sulphonic acids, under conditions where good solutes give peaks a tenth as wide. Forming an ion-pair of this type of molecule has provided a way of overcoming this chromatographic difficulty, as well as giving better detection sensitivity, the more common reason for derivatization[352,119].

An ion-pair is a temporary or reversibly-formed derivative of an ion from the solute molecule. Consider the reversible reaction of benzoate anions with tetra-alkylammonion cations (and ignore the counter-ions for the time being).

In water, because of its high dielectric constant, the two ions will exist separately as solvated benzoate and tetra-alkylammonium species: in non-polar organic solvents they will exist mainly as the ion-pair tetra-alkylammonium benzoate. In a two-phase system where one phase is polar (aqueous) and the other non-polar (organic), the ions and ion-pair will distribute themselves between the two phases. In ion-pair chromatography, one phase (which may be polar or non-polar) is stationary and the other is the mobile phase.

Ion-pair distribution

For the species

S^- solute ion $\qquad$ P^+ pairing ion $\qquad$ SP solute ion-pair

and the equilibrium $S^-_{aq} + P^+_{aq} \leftrightarrows SP_{org}$

the extraction coefficient of the ion-pair is

$$E_{SP} = [SP_{org}] / [S^-_{aq}] [P^+_{aq}]$$

and this can be measured in batch experiments.

The distribution coefficient of the solute is defined by

$$D_S = [SP_{org}] / [S^-_{aq}]$$

and thus, by substitution $D_S = [E_{SP}] [P^+_{aq}]$

If the eluent is buffered with a substantial concentration of P^+ (and counter-ion C^-) then $[P^+_{aq}]$ will be constant, and so therefore will D_S.

If we now consider the counter-ion, the situation is rather more complex. The ion-pair of the added salt CP can, of course, be extracted as ion-pairs into the organic phase so that before any solute is added to the column there is already a distribution of the pairing ion between the two phases.

The equilibria of the components

S^- solute ion $\qquad$ P^+ pairing ion $\qquad$ C^- counter-ion

SP solute ion-pair $\qquad$ CP counter ion-pair

are

ion-pair *extraction* $\qquad C^-_{aq} + P^+_{aq} \rightleftharpoons CP_{org}$

$$S^-_{aq} + P^+_{aq} \rightleftharpoons SP_{org}$$

and ion-pair *exchange* $\qquad S^-_{aq} + CP_{org} \rightleftharpoons C^-_{aq} + SP_{org}$

So the distribution of the solute S can be considered in terms of the ion-pair *extraction* equilibrium, or the ion-pair *exchange* equilibrium, as given below.

ion-pair *extraction* equilibrium

$$E_{SP} = [SP_{org}] / [S^-_{aq}] [P^+_{aq}]$$

$$D_S = [SP_{org}] / [S^-_{aq}]$$

ion-pair *exchange* equilibrium

$$K_{IE} = [C^-_{aq}] [SP_{org}] / [CP_{org}] [S^-_{aq}] = E_{SP}/E_{CP}$$

$$D_S = [SP_{org}] / [S^-_{aq}] = K_{IE} [C^-_{aq}] / [CP_{org}]$$

Which of these equilibria is most appropriate depends on whether it is most practical to control $[P^+_{aq}]$ or $[CP_{org}]$. This will depend on the position of the ion-pair equilibrium below:

counter-ion *extraction* equilibrium

$$C_{aq} + P^+_{aq} \rightleftharpoons CP_{org}$$

$$E_{CP} = [CP_{org}] / [C^-_{aq}] [P^+_{aq}]$$

If most of the salt CP exists mostly as ions in the aqueous (polar) phase, then it is most convenient to control $[P^+_{aq}]$, and apply the equation for ion-pair extraction.

If most of the salt CP exists mostly as the ion-pair in the organic (non-polar) phase, then it is most convenient to control $[CP_{org}]$, and apply the equation for ion-exchange.

Since the stationary phase can either be polar (normal-phase) or non-polar (reversed-phase), and the process can be exchange or extraction (adsorption), there are four ways of separating with ion-pairs, and of course the solute ions can be of either polarity. This classification was given by Knox[304] who gave the following examples as an introduction to 'soap chromatography'[305].

Examples of ion-pair separation processes

ion-pair **partition** chromatography

stationary phase **polar**
aqueous (pH 9)

mobile phase **non-polar**
hexane—Chloroform—pentanol
(75:20:5)

pairing cation P^+_{aq}

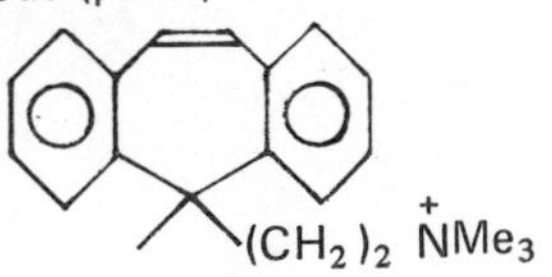

solute anion S^-_{aq}

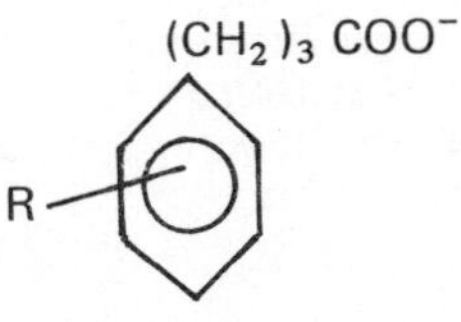

SP_{org}
ion-pair
(UV absorbing)

Schill and others[314]

In this example an aqueous solution of 0.36 M dimethyl protriptylenium was pumped on to a dry column packed with cellulose. Excess aqueous phase was then expelled by pumping the organic eluent. The pairing, and UV absorbing, protritylenium cation was only extracted from the stationary aqueous phase by solute suitable solute anions, for example benzoate or salicylate.

For basic solutes like alkaloids[315] or quaternary ammonium ions[316] picrate is a suitable ion. Knox and Jurand[119] used perchlorate, as have Persson and Karger[317]. This opposite polarity system is shown below.

ion-pair **partition** chromatography

stationary phase **polar**
aqueous (pH 1) $[P^+_{aq}] = 1$ M
on silica

mobile phase **non-polar**
chloroform—butanol
(50:50)

pairing anion P^-_{aq} 2 ClO_4^-

solute cation S^+_{aq}

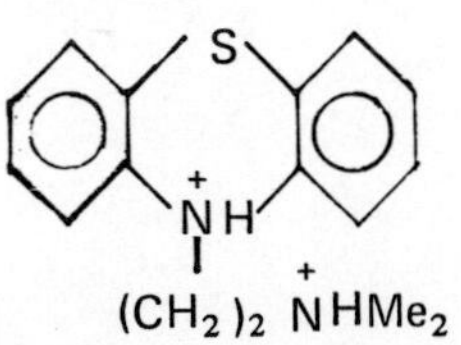

$SP_{2\ org}$

Knox and Jurand[119]

In both of these examples, most of the salt PC exists as ions in the aqueous phase.

The stationary phase can also be made non-polar by coating with liquid and/or bonding, as in the example below.

ion-pair **partition** chromatography — **reversed-phase**

stationary phase **non-polar**
pentanol on ODS-silica

mobile phase **polar**
aqueous $(t\text{-Bu})_4N^+$ $[P^+_{aq}] = 0.03$ M

SP_{org} ⇋ $(t\text{-Bu})_4N^+$ P^+_{aq} ; $R{-}SO_3^-$ S^-_{aq}

(*t*-butylammonium sulphonate)

Schill and others[318]

However for reverse-phase operation it is more usual to operate in the ion-pair *exchange* mode. In the example below, Eksborg, Lagerstrom and Schill[314] used dioctyl hydrogen phosphate (0.1 M) in chloroform as stationary phase. Since this is almost insoluble in the aqueous eluent, nearly all the counter ion pair is in the organic stationary phase.

ion-pair **exchange** chromatography — **reversed-phase**

stationary phase **non-polar**
CP (0.1 M) in chloroform

mobile phase **polar**
aqueous (pH 3.8)
$[C^+_{aq}] = 0.00016$ M

CP_{org} $[CH_3(CH_2)_7]_2P(=O)OH$ ⇅ ephedrine solute S^+_{aq}

SP_{org} $[CH_3(CH_2)_7]_2P(=O)O{-}S$ — H^+ C^+_{aq}

Schill and others[314]

Kraak and Huber[319] coated the stationary phase support with trioctylamine and then passed the eluent of aqueous perchlorate down the column. As the column reached equilibrium with the eluent the trioctylamine reacted to form the perchlorate.

ion pair partition chromatography — **reversed-phase**

stationary phase **non-polar**	mobile phase **polar**
trioctylamine perchlorate	aqueous (pH 2-6) perchlorate
pure CP	$[C^-_{aq}] = 0.05$ M

CP_{org} $[CH_3(CH_2)_7]_3 NHClO_4$

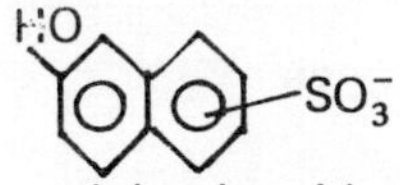

S^-_{aq}

⇅ sulphonic acids

SP_{org} $[CH_3(CH_2)_7]_3$ NH–S ClO_4^- C^-_{aq}

Kraak and Huber[319]

Soap Chromatography of sulphonic acids have been reported by Knox and Laird[305] on surfaces which are both polar (unmodified silica) and non-polar (short alkyl bonded silica) with addition of cetyltrimethylammonium bromide (cetrimide).

They rationalise the interesting observed results as follows:

With a plain silica surface, a small proportion of water in the water–propanol eluent is needed for a reasonable retention. They propose that cetrimide–sulphonate ion-pairs predominate in this propanol-rich system and these ion-pairs are adsorbed by a surface layer of propanol, linked by the polar hydroxyl group to the silica surface, and presenting an essentially non-polar surface. Note that this implies that the original silica surface which was polar is now non-polar, so this is a reversed-phase type mechanism for adsorption.

With the reverse-phase 'short alkyl' bonded silica a much more aqueous system was needed. The surface is also non-polar because of an associated layer of propanol molecules, but a layer of cetrimide cations will also lie between the non-polar surface and the now highly polar aqueous eluent. The alkyl part of the molecule will interact with the alkyl coating of the silica, and the polar nitrogen end will be solvated by water. Adsorption of sulphonic acids takes place on these groups.

Separation of Anthraquinone sulphonic acids by different methods

Silica

(Spherisorb S5W, 0.005M sulphuric acid in acetonitrile)

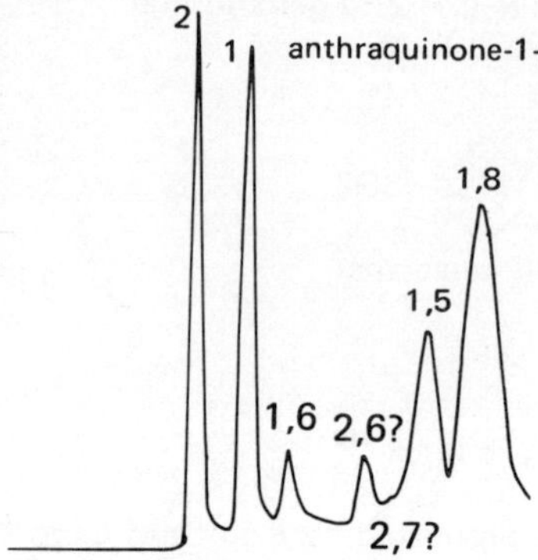

Ion-pair Derivative on Silica

(Spherisorb S5W, acetonitrile, water and cetrimide 95:5:2)

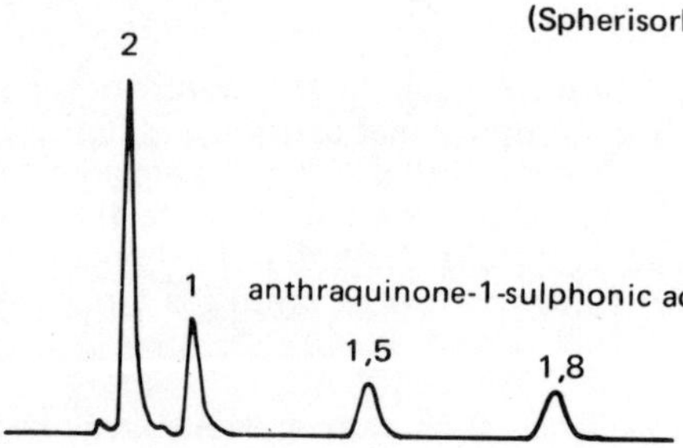

Ion-pair Derivative on ODS-Silica

(S5W-ODS, acetonitrile, water and cetrimide 50:50:2)

1 anthraquinone-1-sulphonic acid

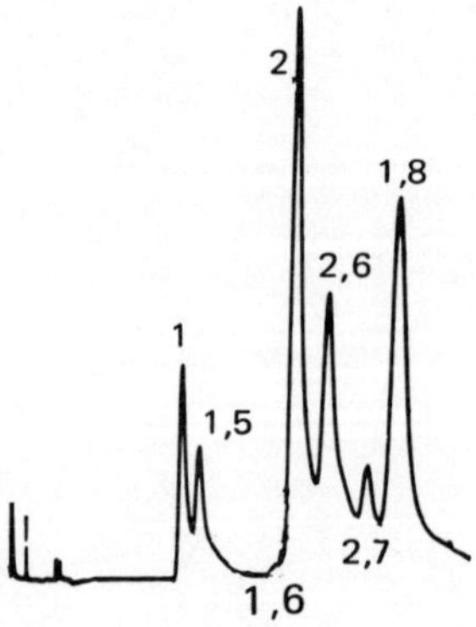

Source: Ms A Hunter, ICI Organics Division
see also Knox and Laird[305]

Post-column derivatization

summary • **requires careful choice of hardware and reagents**
• **development needed to use very efficient columns**

The pioneers of liquid chromatography, Stein and Moore[167] also pioneered post-column reaction detection for their analysis of amino-acids.

Compared to pre-column derivatization:

- formation of artefacts is less likely
- simultaneous dual detectors can be used
- choice of eluent may conflict with reagents, because of insolubility or reaction
- reaction must be very fast (<30 s)
- the reagent itself must not be detectable
- detectable species may decompose and decay

As column efficiencies have improved, both in speed and resolution, the technological problems of post-column derivatization have increased. Columns packed with micro-particle silicas or ion-exchange resins (or pellicular materials) can produce peaks at intervals of 1–100 s. The detection system must be able to operate at this sort of rate or more band broadening will be caused by operating the column too slowly (say at reduced velocities <1, see section 3). Peak volumes have also fallen, though they could be increased somewhat by using larger diameter columns, say 3/8 in diameter, 8 mm bore instead of the 1/4 in size now most common, which would increase volumes about threefold without causing any loss of resolution. (In practice, this diameter has been found to give slightly higher efficiency than ~5 mm bore columns.) But more solvent and reagent will be used for each analysis.

Users of post-column detection are now tackling the problems posed by high speed and small peak volumes, and some examples are tabulated below. Design of the tee and reaction coil are critical areas to avoid losing resolution from tubes that are too long or too large diameter (see section 12.4) but with time for complete reaction, but not too long to avoid decay. Coils of PTFE with bores of 0.2 to 0.5 mm and length of up to 10 m seem suitable.

Typical scheme for post-column derivatization

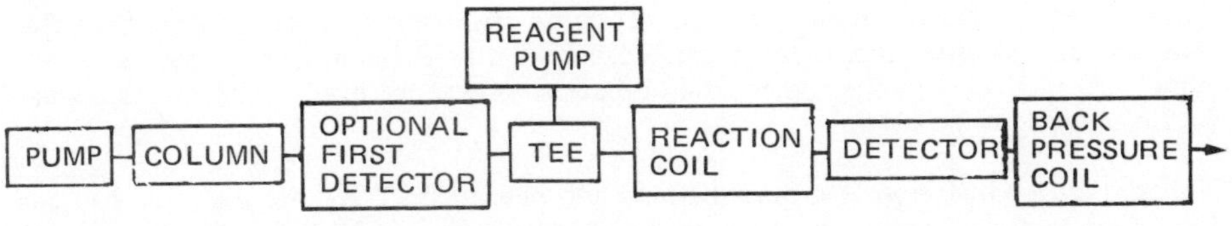

8.11 Ultimate Limits on Performance

summary
- **small performance gains for small particles (<3 μm) or longer columns (>1 m)**
- **selective packings and eluents for greater separating power**
- **gradient elution can increase peak capacity**

As the performance of liquid chromatography has improved so have the demands of analytical problems. An important aim of theory must be to guide the practical search for still higher performance. Should we use higher pressures, longer columns or yet smaller particles? Briefly the answer is no, for columns that have been described are already on the boundary beyond which there are sharply diminishing returns[248,285].

The main potential for increased separating power lies in more selective packing materials and solvent systems. Gradient elution is the obvious way to increase the peak capacity.

Speed

summary
- **small particles for fast analyses**
- **practical limit on speed from detector volume, mass sensitivity and response speed**

If we compare columns with the *same reduced length* operated under the same conditions and at the *same reduced velocity,* the particle size is seen as the most important factor determining speed. (Since the form of the reduced plate height—reduced velocity curve is the same regardless of size, any speed increase, at the expense of plate height, can be achieved by operating at a higher reduced velocity for any particle size.)

Substituting $l\, d_p$ for L and rearranging the equation $\nu = L\, d_p / t_o\, D_m$

we obtain $t_o = l\, d_p^2 / \nu\, D_m$

thus for comparable colums $t_o \propto d_p^2$

so smaller particles give much faster analyses. (Incidentally, it also shows a small advantage from less viscous solvents which give higher diffusivity, as well as lower pressure drop.)

Thus limits on the ultimate *speed* of LC come from the smallest volume that the detector can sample, the mass it can detect in that volume, and its speed of response. It is this, together with the injection problem, that limits the smallest particles that can be used.

This detector limitation is a practical one and new technology, for example sources like lasers and smaller photodetectors, may lead to some small improvements.

Ultimate Separating Power

summary **• limit on plate number from increase in viscosity and heat generated by very high pressures**

There is now some evidence[140] that there are maximum pressure limits which are rather more fundamental. There are two factors:

- the increase in viscosity (and consequent reduction in solute diffusivity) caused by the applied pressure which reduces chromatographic retention and speed of analysis.
- heat generated as the liquid is forced through the column sets up radial (and axial) temperature gradients whose effects are complex but are almost certain to be bad chromatographically. This is worst for wide bore columns.

This pressure limit is *not* imposed by the pump but by chromatography.

It is conceivable that beyond the boundaries imposed by these effects there are turbulent flow conditions (reference 1, p. 217) that will offer yet higher performance, but results and predictions so far are not very encouraging. Pressure increases will have to be some orders of magnitude[221,222] higher to get into a region where plate heights are reducing and rises in eluent viscosity may make the application of extreme pressures self-defeating.

Pressure in the region of 100 to 400 bar (200–10 000 psi) and above will have an effect on chromatography[140]. The following equations, based on reduced parameters (and corresponding to those in absolute terms by Halasz et al[140]) show some implications of this pressure limit. Since so little is known about the effect of very high pressures, these calculations are intended only to give a feel for the limiting values.

Combining equations for the reduced parameters ϕ, ν and l gives

$$P = \phi \eta D_m \nu l/d_p^2$$

If the following values are assumed:

$$\phi = 500,\ \eta = 3 \times 10^{-4}\ \mathrm{N\,s/m^2},\ D_m = 3 \times 10^{-9}\ \mathrm{m^2/s},\ \nu = 5\ \text{(for minimum h)}$$

then the more interesting relationship between P, d_p and l can be explored. (Naturally an eluent with as low a viscosity as possible should be selected.)

Thus $P = 2.25 \times 10^{-9}\ l/d_p^2$

If we set $P_{max} = 200\ \text{bar} = 2 \times 10^7$ Pa, then

$$d_p^2 = \frac{2.25 \times 10^{-9}\ l}{2 \times 10^7} = 1.1 \times 10^{-16}\ l$$

and $d_p = 1.05 \times 10^{-8} \sqrt{l}$ or $d_p\ (\mu m) \simeq 10^{-2} \sqrt{l}$ and $l \simeq 10^4\ d_p^2\ (\mu m)$

This relationship can be used in several ways.

If a column of reduced length 20 000 is required (typical of 200 mm columns packed with 10 μm particles) then the smallest particle size that could be used to speed up the analysis is $10^{-2} \times \sqrt{2 \times 10^4} = 1.4\ \mu$m.

Its actual length can be calculated from $L = l\, d_p = 2 \times 10^4 \times 1.4 \times 10^6 = 28$ mm.

This is also the longest column that can be packed with 1.4 μm particles before pressure induced problems arise.

If we relax the pressure limit to 800 bar, an increase of 4, the pressure-induced effects will certainly intrude, and consider an 'ultimate plate number' column[174] with $d_p = 3\ \mu$m then

$$l = \frac{P\, d_p^2}{2.25 \times 10^{-9}} = \frac{8 \times 10^7 \times 9 \times 10^{-12}}{2.25 \times 10^{-9}} = 300\,000$$

which can only be ~1 m long. At best we can expect h = 2 and from N = l/h, a plate number of 150 000.

The analysis time can be assessed from

$$t_o = l\, d_p^2 / \nu\, D_m = \frac{3 \times 10^5 \times 9 \times 10^{-9}}{5 \times 3 \times 10^{-9}} = 190 \text{ s.}$$

Paradoxically to make columns with higher plate numbers it is necessary to use *larger* particles, the method used by Scott[385] (holder of the plate number record at the time of writing).

If we demand a reduced length of L of 1 000 000 and we use 10 μm particles, then the column will need to be $10^6 \times 10 \times 10^{-6} = 10$ m long,

and the pressure $P = 2.25 \times 10^{-9} \times 10^6 / 10^{-10} = 2.25 \times 10^7$ Pa = 225 bar.

In this case the column cannot operate in the infinite diameter mode and a larger reduced height may have to be accepted.

If h = 4, then $N = 10^6/4 = 250\,000$

and the time $t_o = 10 \times 10^{-5} / 5 \times 3 \times 10^{-9} = 6\,700$ s.

By using a higher pressure this may be reduced to near 1 000 s but the price of about doubling the plate number (only a 50% increase in resolution) is about a fiftyfold increase in analysis time compared to the 3 μm column described above.

These two examples illustrate the boundary regions of column performance. Certainly Giddings[223] estimate of $N \approx 10^8$ seems optimistic if these equations are applied. If $N = 10^8$ and $h = 2$, $l = 2 \times 10^8$, and $P = 200$ bar, then

$$d_p\,(\text{min}) = 10^{-2} \times 1.4 \times 10^4 = 140\ \mu\text{m},$$

which implies $L = 2 \times 10^8 \times 140 \times 10^{-6} = 28\,000$ m, a rather long column. Analysis times are also inconveniently long

$$t_o = 2.8 \times 10^4 \times 140 \times 10^{-6}/5 \times 3 \times 10^{-9} = 2.6 \times 10^8\ \text{s} \approx 8\ \text{yr}$$

and one chromatogram a lifetime's work!

summary • **1 m x 3/8 diameter with 5 μm packing most practical column for ultimate separating power**

In practice then a column around 1 m packed with commercially available spherical (for lower pressure drop) 5 μm packing material and operated at 200 bar seems the best compromise for the 'ultimate column' in LC.

For this column $l = 2 \times 10^7 \times 25 \times 10^{-12}/2.25 \times 10^{-9} = 200\,000$,

and if the column diameter is large enough to permit operation in the infinite diameter mode we may expect 100 000 plates under ideal conditions. (>7 mm is calculated to be needed for 100 μl injections, so a 3/8 diameter column is suitable). Since

$$t_o = 1 \times 5 \times 10^{-12}/5 \times 3 \times 10^{-9} \approx 300\ \text{s}$$

a practical analysis time of about 1 hour can be expected.

A chromatogram showing a column of this configuration under test conditions is shown below.

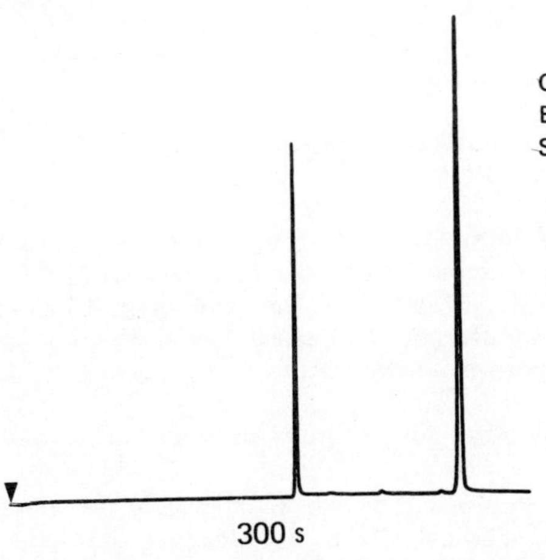

Column: Spherisorb S5W, 1m x 8 mm bore
Eluent: hexane:acetonitrile 99:1
Solutes: xylene and nitrobenzene

9 INFORMATION ON LIQUID CHROMATOGRAPHY

9.1 General sources

LC has developed so rapidly over the past decade that information has quickly become outdated: so will some of the content of this book. Thus a list of manufacturers' products and addresses has not been included because it will too quickly be out of date.

Classical LC methods are well documented: Lederer and Lederer[11] is a typically comprehensive book. The literature up to 1969 has been comprehensively reviewed[130]. This information is by no means obsolete though it is not always obvious how to apply information on separations on classical materials to modern microparticles.

Books

Books are an important source though they tend not to be very up to date. A number are listed in section 9.4, and several others are in preparation. That of Kirkland and Snyder[3] stands out as an authoritative and comprehensive text on modern LC.

Reviews

Reviews of instrumentation, theories and applications are more often up to date because they are produced and printed more quickly. Applications are particularly comprehensively reviewed in the biennial Review Section of Analytical Chemistry[132], and in the Bibliography Section on Column Chromatography[131] published four times each year in Journal of Chromatography. (References given are the most recent at the time of writing.)

Primary publications

Most of the primary publications about LC are in:

Analytical Chemistry (American Chemical Society)
Journal of Chromatography (Elsevier)
Journal of Chromatographic Science (Preston Technical Abstracts Co)
Chromatographia (Friedr. Vieweg & Sohn/Pergamon)

Analytical Chemistry publishes many interesting advertisements and instrument reviews in the A section pages, and an annual list of manufacturers' products and addresses (in the US) in their Laboratory Guide[133]. The Journal of Chromatographic Science also tabulates all products and addresses[134]. All these journals describe new products based on manufacturer's information.

Some papers are also now appearing in other subject journals where LC is more of a means than an end, for example:

Analytical Biochemistry, Journal of Polymer Science (particularly permeation chromatography), Separation Science, Clinical Chemistry, Journal of Medicinal Chemistry, Journal of Pharmaceutical Science.

The following abstracts specifically of the LC literature are produced:

Gas and Liquid Chromatography Abstracts, sponsored by the Chromatography Discussion Group and published by Applied Science Publishers, Ripple Road, Barking, Essex, UK (4 issues per year)

LC Abstracts, published by Preston Technical Abstracts, PO Box 312, Niles, Illinois, USA 60648 (6 issues per year)

PRM Science and Technology Agency, 787 High Road, North Finchley, London N12, UK (4 issues per year)

Manufacturers information

Instrument manufacturers produce a surprising amount of data on applications, as well as some of the expected data on their instrumentation. Their applications laboratories often see a very wide range of samples so they are experienced, if not impartial. Applications notes are produced by at least the following:

Applied Science, Du Pont, Pharmacia, Pye Unicam, Perkin-Elmer, Varian, Waters Associates, Spectra Physics.

Symposia

Symposia reports are often specially topical. Many are published as special issues of the Journal of Chromatography and the Journal of Chromatographic Science, often additionally in book form.

Unpublished information

This mine of information probably exceeds all the sources above in quantity. Its quality may be more uneven (though not all the primary literature is above contention) but it is highly inaccessible except by personal contact.

Abstracts

Chemical Abstracts (mainly sections 79 and 80) is a useful source and digest of information in literature not primarily for chromatographers, particularly patents. The growing number of publications not in English (particularly in Japanese in the last year or two), are often abstracted in English.

UKCIS, Nottingham, UK produce a monthly macroprofile listing all papers on LC from the Chemical Abstracts tapes. Some companies like ICI search these tapes against their own profiles to selectively disseminate information (SDI) in card form. Design of effective profiles for LC is not easy because of the wide variety of keywords which are used by authors. The rapid growth of papers on applications has lead to a rapid increase in the number of references resulting from these profiles and this can be expected to continue steadily.

9.2 Bibliography

1 J.C. Giddings, *Dynamics of Chromatography,* Pt. I, E. Arnold (London) and M. Dekker (New York), 1965, ISBN 8247 1225 0.

2 A. Zlatkis, Ed., *Advances in Chromatography (Las Vegas Symposium)* Preston Technical Abstracts, Evanston, Illinois, 1965.

3 L.R. Snyder and J.J. Kirkland, *Introduction to Modern Liquid Chromatography,* J. Wiley, New York, 1974, ISBN 0 471 81019 3.

4 J.N. Done, J.H. Knox, J. Loheac, *Applications of High-speed Liquid Chromatography,* J. Wiley, London, 1974, ISBN 0 471 21784 0.

5 J.J. Kirkland, Ed., *Modern Practice of Liquid Chromatography,* J. Wiley, New York, 1971, ISBN 0 47 48878 X.

6 P.R. Brown, *High Pressure Liquid Chromatography: Biochemical and Biomedical Applications,* Academic Press, New York, 1973, ISBN 12 136950 1.

7 N. Hadden et al, *Basic Liquid Chromatography,* Varian Associates, Walnut Creek, Calif., 1972.

8 L.R. Snyder, *Principles of Adsorption Chromatography,* M. Dekker, New York, 1967.

9 E. Stahl, *Thin-layer Chromatography: a Laboratory Handbook, 2nd Ed.,* Academic Press, New York, 1969, ISBN 04 547 001 4.

10 E. Grushka, Ed., *Bonded Stationary Phases in Chromatography,* Ann Arbor Science Publishers, 1975, ISBN 250 40039 1.

11 E. Lederer and M. Lederer, *Chromatography,* Elsevier, London, 1957, ISBN 444 40359 0.

12 S.G. Perry, R. Amos and P.I. Brewer, *Practical Liquid Chromatography,* Plenum Press, New York, 1972, ISBN 306 30548 8.

13 H. Determann, *Gel Chromatography: Gel Filtration, Gel Permeation, Molecular Sieves,* Springer Verlag, New York, 1968.

14 L. Fischer, *Introduction to Gel Chromatography,* North-Holland, London, 1969.

15 O. Samuelson, *Ion Exchange Separations in Analytical Chemistry,* J. Wiley, New York, 1963.

16 W. Reiman and H.F. Walton, *Ion Exchange in Analytical Chemistry,* Pergamon, New York, 1970.

17 A.B. Littlewood, *Gas Chromatography,* Academic Press, New York, 1970, ISBN 0 12 452756 6.

18 S.G. Perry, Ed., *Gas Chromatography 1972,* Applied Science, London, 1973, ISBN 85334 479 5.

19 O. Lauer, *Grain Size Measurements on Commercial Powders,* issued by Alpine AG, Ausburg, Germany (1966)

20 J.A. Riddick and W.B. Bunger in *Organic Solvents, Physical Properties and Purification,* Ed. A. Weissberger, Wiley-Interscience, 3rd Ed, 1970
ISBN 0471 92726 0

21 T. Allen, *Particle Size Measurement,* Chapman and Hall, (1975)
ISBN 412 13490 X

22 J.X. Khym, *Anal. Ion-Exchange Procedures in Chem. Biol.,* Prentice-Hall, (1974)
ISBN 013 034942 9

25 H.H. Weetall, Ed., *Immobilized Enzymes, Antigens, Antibodies, and Peptides, Preparation and Characterization,* Dekker, NY (1975)

26 A.V. Kiselev and Y.I. Yashin, *Gas-Adsorption Chromatography,* Plenum, NY (1969)

review adsorbents

23 D.D. Bly in *Physical Methods in Macromolecular Chemistry,* ed. B. Carroll, Vol. 2, p.1-90, Dekker, N.Y. (1972). ISBN 0 8247 1085 1

24 B.G. Linsen, ed., *Physical and Chemical Aspects of Adsorbents and Catalysts,* Academic Press, London (1970) ISBN 12 451150 3.

9.3 References

References are chosen frugally on the following criteria:

the first suggestion or realisation of a new concept;
a lucid presentation for the practical user;
a summary of what seems a definitive conclusion;
a current review;
or the most recent report which will refer to earlier work.

This list is not intended as a comprehensive list of all LC references: the great body of original work is acknowledged by citation elsewhere.

29 W. Shakespeare, *Hamlet,* **2** (2) (1603)

31 M. Martin, G.Blu, C.Eon, G. Guiochon, *J. Chromatog.,* **112**, 399 (1975) **large piston pumps**

32 K. Keck, *Anal. Biochem.,* 39, 288 (1971) **gradient and mixing system**

33 K. Keck, (LKB Productur AB, Stockholm) *German Pat.* 1 598 683 (1972) **Ultrograd**

34 F.R. MacDonald, *International Lab.,* 39. July/August (1973) **power mixing**

35 M.S. Denton, T.P. DeAngelis, A.M. Yacynych, W.R. Heinemann, T.W. Gilbert, *Anal. Chem.,* **48**(1), 20 (1976) **rapid scanning UV detector**

36 D.G.I. Kingston and B.B. Gerhart, *J. Chromatog.,* 182 (197

37 R.E. Majors and M.L. Hopper, *J. Chrom. Sci.,* **12**, 767 (1974)

38 A. Pryde, *J. Chrom. Sci.,* **12**, 486 (1974) **review bonded phases**

39 P.A. Bristow, *Anal. Chem.,* **48**, 237 (1976) **gradient by time-proportioning**

40 A.J.P. Martin and R.L.M. Synge, *Biochem. J.,* **35**, 1358 (1941) **chromatography**

41 J.C. Gidding, *Anal. Chem.,* **35**, 1338 (1963)

42 C. Horvath, B. Preiss, S.R. Lipsky, *Anal. Chem.,* **39**, 1422 (1967) **pelliculars**

43 I. Halasz and C. Horvath, *US Patent* 3,340,085; *British Patent* 1,016, 635 **pelliculars**

44 J.N. Done, G.J. Kennedy, J.H. Knox, ref. 18 p. 145

45 C.R. Wilke and P. Chang, *Am. Inst. Chem. Eng. J.,* **1**, 264 (1955) **diffusion coefficients**

46 E.V. Piel, *Anal. Chem.,* **38**, 670 (1966) **colloidal particles for LC**

47 K.J. Bombaugh, *J. Chromatog.,* **107**, 201 (1975) **gradient system**

48 D. De Kee and H. Laudie, *Hydrocarbon Process.,* 53(9), 224 (1974) **diffusion coefficients**

49 P. Chang and C.R. Wilke, *J. Phys. Chem.,* **59**, 592 (1955) **diffusion coefficients**

50 W. Hajduck and S.C. Cheng., *Chem. Eng. Sci.,* **26**, 635 (1971) **diffusion coefficients**

51 A. Einstein, *Z. Electrochem.,* **14**, 1908 (1908)

52 J.H. Knox and P.A. Bristow, submitted to *Chromatographia.* **column testing**

53 W.R. Gambill, *Chem. Engr.,* June 30, p. 113 (1958)

54 Deutsche Normen DIN 2093, Disc Springs—Dimensions and Quality Properties, (1967) (English translation) **disc springs**

55 Deutsche Normen DIN 2092, Design sheet on Disc Springs (1967)

56 Eng. Mat. Design **19**, 13 (1975) **stainless steels**

57 Nickel Stainless Steels—notes on fabrication and machining, International Nickel Ltd, Thames House, Millbank, London

58 J.L. Meek, *Anal. Chem.,* **48**(2), 375 (1976) **constant pressure pump**

59 H.W. Johnson, V.A. Campanile, H.A. LeFre, *Anal. Chem.,* **39**, 32 (1967)

60 R.L. Anderson, L.A. Guildner, R.E. Edsinger, *Rev. Sci. Inst.,* **41**, 1076 (1970)

61 S. Sherratt in *Encyclopedia of Chemical Technology,* 2nd edition **9**, Ed. Kirk-Othner, Wiley-Interscience, New York, 1966, p. 805-831

62 D.M. Wells, *Engineering,* **211**, 1029 (1971)

63 ICI Plastics Division, Welwyn Garden City, UK, *Technical Service Note* F12

64 ditto F10

65 ditto F13

66 ditto FTD130

67 3M UK Ltd, Wigmore St, London, *Technical information on Kel-F Plastics, Physical Properties and Processing,* 1976

68 ICI Plastics Division, Welwyn Garden City, UK, Technical bulletin PPTD537

69 ditto PMC103

70 ditto PMC104

71 ditto N110

72 ditto N.TD5

73 Amcel Ltd, (Celanese Corp in USA), Watford, UK. Technical sheet on Selection of Engineering Resins, 1975
74 H. Forestier and L. Truffert, *Analusis,* **3**, 271 (1975) **toxicity and explosion hazards**
75 R.L. Redding, *Intrinsic Safety,* McGraw, London, 1971, ISBN 07 094224 2
76 L.A. Pachla and P.T. Kissinger, *Anal. Chem.,* **48**, 364 (1976) **electrochem detector**
77 ASTM E387-69T, Estimating Stray Radiant Energy (1969)
78 E.S. Watson, U.S. Pat. 3591801 (1971); C.A. 75: P94386d **UV detector**
79 *Handbook of Chemistry and Physics,* ed. R.C. Weast, 56th Ed, CRC Press, Cleveland, Ohio, 1975, ISBN 0 87819 455 X
80 A. Bylina, D. Sybilska, Z.R. Grabowski, J. Koszewski, *J. Chromatog.,* **83**, 357 (1973)
81 D.A. Yates and T. Kuwana, *Anal. Chem.,* **48**, 510 (1976) **rapid-scan UV detector**
82 N.K. Freeman, F.T. Upham, A.A. Windsor, *Anal. Letters,* **6**, 943 (1973) **laser IR**
83 C.D. Carr, *Anal. Chem.,* **46**, 743 (1974)
84 J.W. Strojek, G.A. Gruver, T. Kuwana, *Anal. Chem.,* **41**, 481 (1969)
85 J.Y. Chuang and J.F. Johnson, *Sep. Sci.,* **10**, 161 (1975) **step gradient,derivatives**
86 Y. Tatmi, R. Crosmun, N.M. Larson, *Anal. Chem.,* **48**, 326 (1976) **light source noise**
87 N.C. Fawcett and R.D. Caton, *Anal. Chem.,* **48**, 600 (1976) **determination of diffusion coefficients**
88 C.L. Blank, *J. Chromatog.,* **117**, 35 (1976) **dual electrochem detector**
89 P.T. Kissinger, C. Refshauge, R. Dreiling and R.N. Adams, *Anal. Letters,* **6**, 465 (1973)
90 J.G. Koen, J.F.K. Huber, H. Poppe and G. den Boef, *J. Chrom. Sci.,* **8**, 192 (1970)
91 C.J. Asher and R.J. Prescott, *Anal. Biochem.,* **69**, 618 (1972) **flowcell for ^{75}Se**
92 F.W. Karasek, *Research and Development,* **26**, 34 (1975) **review novel detectors**
93 R.J. Dolphin, F.W. Willmott, A.D. Mills and L.P.J. Hoogeveen, *J. Chromatog.,* in the press
94 G.B. Sieswerda and H.L. Polak, *J. Radioanal. Chem.,* **11**, 49 (1972) **scintillators**
95 R.F. Borch, *Anal. Chem.,* **47**, 2437 (1975) **step gradient**
96 E.S. Watson, *U.S. Pat.* 3 386 332 (1968); C.A. 72: 86007d **refractometer**
97 J.L. Water, *U.S. Pat.* 3 522 725 (1970); C.A. 73: P105224m **chromatograph and refractometer**
98 D.R. Jones, H.C. Tung and S.E. Manahan, *Anal. Chem.,* **48**, 7 (1976) **AAS**
99 M.J. Holdoway, Atomic Energy Res. Estab., Harwell, UK, M2749 (1976) ISBN 0 70 580416 X **physical properties of absorbents**
100 R. Delley, *Chromatographia,* **9**, 10 (1976) **'efficiency' in chromatography**
101 R.E. Dessy, W.G. Nunn, C.A.Titus, W.R. Reynolds, *J. Chrom. Sci.,* **14**, 195 (1976) **fast scan detector**
102 H.B. Peterson, *J. Chrom. Sci.,* **14**, 211 (1976) **electropolishing columns**
103 J.H. Knox, *Notes for LC Workshop,* ICI Pharmaceutical Division, Macclesfield, UK, (1975)
104 J.H. Knox, J.F.K. Huber, L.R. Snyder, *J. Chrom. Sci.,* **13**, 25 (1972) **comparative theories of LC**
105 P.B. Hamilton, D.C. Bogue, R.A. Andersen, *Anal. Chem.,* **32**, 1782 (1966) **small ion-exchange resins**
106 J.H. Knox and J.F. Parcher, *Anal. Chem.,* **41**, 1599 (1969) **infinite diameter effect**
107 G. Deininger and I. Halasz, *J. Chrom. Sci.,* **9**, 83 (1971) **detector broadening**
108 J.F.K. Huber, *J. Chrom. Sci.,* **7**, 172 (1969) **detector evaluation**
109 G.R. Laird, J. Jurand and J.H. Knox, *Proc. Soc. Analyt. Chem.,* 311 (1974) **microparticles**
110 G. Brooker, *Anal. Chem.,* **43**, 1095 (1971) **reducing detector flow sensitivity**
111 J.R. Felton, Du Pont, *U.S.P.* 3 614 452 (1971) **T-cell detector flow sensitivity**
112 J.D. Swain, *Process Eng.,* 100, Nov (1965) **electropolishing**
113 I. Halasz and J. Asshauer, *J. Chrom. Sci.,* **12**, 139 (1974) **drilled columns**
114 J.J. Kirkland, *J. Chrom. Sci.,* **7**, 361 (1969)
115 R.E. Majors, Symposium on Advances in Chromatog, Miami, USA (1970)
116 L.H. Thacker, W.W. Pitt, S. Katz, C.D. Scott, *Clin. Chem.,* **16**, 626 (1970) **254 nm detector**
117 R.J. Laub, *Res. and Dev.,* **25**, 24 (1974) **effect of column surface**
118 D. Horne, J.H. Knox, L. McLaren, *Separ. Sci.,* **1**, 531 (1966)
119 J.H. Knox and J. Jurand, *J. Chromatog.,* **110**, 103 (1975) **tetracycline**
120 J.H. Knox, G.R. Laird and P.A. Raven, *J. Chromatog.,* **122**, 129 (1976) **radial dispersion**

121 Amer. Petrol. Inst., API RP27, 3rd Ed (1952) **permeability**
122 M. Walter and L. Ramaley, *Anal. Chem.*, **45**, 165 (1973) **purification of CH_3CN**
123 R. J. Dolphin and F. W. Willmott, *J. Chrom. Sci.*, **14**, 584 (1976) **column switching**
124 J.H. Ross and R.L. Shank in "Polymer Molecular Weight Methods", *Adv. in Chem.*, Series 125, p 108 (1973)
125 L.R. Snyder, *Anal. Chem.*, **46**, 1384 (1974) **solvent in LSC review**
126 L.R. Snyder, *J. Chromatog.*, **92**, 223 (1974) **solvent properties**
127 D.L. Saunders, *Anal. Chem.*, **46**, 470 (1974) **solvent selection**
128 A.F.M. Barton, *Chem. Rev.*, **75**, 748 (1975) **solubility parameters**
129 P.R. Young and H.M. McNair, *Anal. Chem.*, **47**, 757 (1975) **amines**
130 Literature review up to 1969, *J. Chrom. Sci.*, **8**, D2 (July 1970)
131 Biblio. Section on Column Chromatography, *J. Chromatog.*, **118**, B18 (1976)
132 H. Walton, *Anal. Chem.*, **48**(5), 52R (1976); G. Zweig and J. Sherma, *Anal. Chem.*, **46**(5), 73R (1974)
133 Laboratory Guide, *Anal. Chem.*, **47**, 63 and 64 (1975)
134 Product Listing and Company Directory, *J. Chrom. Sci.*, **14**, 71 and 107 (1976)
135 L. Rohrschneider, *Anal. Chem.*, **45**, 1241 (1973) **solvent theory**
136 J.F.K. Huber, E.T. Alderlieste, H. Harren, and H. Poppe, *Anal. Chem.*, **45**, 1337 (1973) **solvent theory**
137 H. Colin, C. Eon, G. Guiochon, *J. Chromatog.*, **119**, 41 (1976) **carbon**
138 R.P.W. Scott and P. Kucera, *J. Chromatog.*, **119**, 467 (1976)
139 ASTM E388-72, Spectral Bandwidth and Wavelength Accuracy of Fluorescence Spectrometers
140 I. Halasz, R. Endele, J. Asshauer, *J. Chromatog.*, **112**, 37 (1975) **ultimate limits**
141 H. Engelhardt and H. Elgass, *J. Chromatog.*, **112**, 415 (1975) **gradient problems**
142 W.E. Bambrick and J.T. Geohegan, *J. Chrom. Sci.*, **8**, 232 (1970)
143 H. Veening, CRC Crit. Rev., *Anal. Chem.*, **5**, 165 (1975) **review**
144 R.L. Pecsok and D.L. Saunders, *Anal. Chem.*, **40**, 1756 (1968)
145 S. Katz and L.H. Thacker, *J. Chromatog.*, **64**, 247 (1972)
146 D. Berek, T. Blehs, Z. Pevna, *J. Chrom. Sci.*, **14**, 560 (1976) **solvent peaks**
147 G.H. Brilmyer, S.C. Lamey, J.T. Maloy, *Anal. Chem.*, **47**, 2304 (1975) **electrochem. detector**
148 J. Burns and B.F. Williamson, personal communication
149 R.P.W. Scott, *British Pat.* 998 107 (1965); *U.S. Pat.* 3 292 420 (1966)
150 R.P.W. Scott and J.G. Lawrence, *J. Chrom. Sci.*, **8**, 65 (1970)
151 J.J. Szakasits and R.E. Robinson, *Anal. Chem.*, **46**, 1648 (1974); *U.S. Pat.* 3 788 479 (1974) **alumina disc—FID detector**
152 P.R. Jones and S.K. Yang, *Anal. Chem.*, **47**, 1000 (1975) **silicone LC-MS**
153 F.W. McLafferty, R. Knutti, R. Venkataragharan, P.J. Arpino, B.G. Dawkins, *Anal. Chem.*, **47**, 1503 (1975) **LC-MS**
154 D.I. Carroll, I. Dzidic, R.N. Stillwell, M.G. Horning, E.C. Horning, *Anal. Chem.*, **46**, 706 (1974) **LC-MS**
155 L. Schutte, *J. Chromatog.*, **72**, 303 (1972) **radioactivity monitoring**
156 B. Fleet and C.J. Little, *J. Chrom. Sci.*, **12**, 747 (1975) **electrochem. detector**
157 P.B. Hamilton, D.C. Bogue, R.A. Ánderson, *Anal. Chem.*, **32**, 1782 (1960)
158 E. Godbille and P. Devaux, *J. Chrom. Sci.*, **12**, 564 (1974) **prep LC**
159 M. Bakken and V.I. Stenberg, *J. Chrom. Sci.*, **9**, 603 (1971) **interferometer detector**
160 B.G. Julin, H.W. Vandenbom, J.J. Kirkland, *J. Chromatog.*, **112**, 443 (1975) **S and P detectors**
161 J.N. Little and G.J. Fallick, *J. Chromatog.*, **112**, 389 (1975) **taper flowcell**
162 R.C. Buchta and L.J. Papa, *J. Chrom. Sci.*, **14**, 213 (1976) **electrochem. detector**
163 H.R. Linder, H.P. Keller and R.W. Frei, *J. Chrom. Sci.*, **14**, 234 (1976) **slurry packing**
164 R.M. Riggin, A.L. Schmidt, *J. Pharm. Sci.*, **64**, 680 (1975) **electrochemical detector**
165 L.S. Ettre and C. Horvath, *Anal. Chem.*, **47**(4), 422A (1975) **LC history**
166 M. Tswett, *Trav. Soc. Nat. Warsovie*, **14**, (1903) English translation obtainable from W. Woelm, Eschwege, W. Germany)
167 D.H. Spackman, W.H. Stein, S. Moore, *Anal. Chem.*, **30**, 1190 (1958) **ninhydrin**
168 E.S. Watson, personal communication
169 J.I. Hollies, personal communication
170 S. Udenfriend, S. Stein, P. Böhlen, W. Dairman, W. Leimgruber, M. Weigele, *Science*, **178**, 871 (1972) **application of fluorescamine**
171 S. Katz and W.W. Pitt, *Anal. Letters*, **5**, 177 (1972)

172 J.V. Dawkins and M. Hemming, *J. Appl. Poly. Sci.*, **19**, 3107 (1975) **IR detector for PC**
173 J.V. Dawkins, in *Indust. Poly. Char. by Mol. Wt.*, Ed. J.H.S. Green and R. Dietz, Transcripta Books, London (1973), p. 51 **PC symposium**
174 P.A. Bristow, P.N. Brittain, C.M. Riley and B.F. Williamson, *J. Chromatog.*, **131**, 57 (1977) **slurry packing**
175 J.F. Norris, PhD Thesis, University of Birmingham (1972)
176 Phillips, *BP* 844 619 (1960); 820 943 (1959) **phosphors**
177 Recommendations on Nomenclature for Chromatography, *Pure and Appl. Chem.*, **37**, 447 (1974) **nomenclature**
178 J.J. Kirkland, *J. Chrom. Sci.*, **10**, 593 (1972) **silica for fast PC**
179 K.K. Unger, R. Kern, M.C. Ninou, K-F. Krebs, *J. Chromatog.*, **99**, 435 (1974) **silica for PC**
180 K.J. Bombaugh and R.F. Lerangie, *Anal. Chem.*, **41**, 1359 (1969) **fast PC**
181 N. Baba and S. Sato, *Japan Anal.*, **20**, 208 (1971), C.A. 72: 70939y and 75: 6491f **PC with alumina**
182 A. Titterton, in ref. 173, p. 83 **PC with silica**
183 L.R. Whitlock and R.S. Porter, *J. Chrom. Sci.*, **10**, 437 (1972)
184 M.F. Vaughan in ref. 173, p. 111 **prep PC**
185 D.D. Bly, H.J. Stoklosa, J.J. Kirkland, W.W. Yau, *Anal. Chem.*, **47**, 1810, and correction p. 2328 (1975) **flowrate errors in PC**
186 Yu. A. Eltekov and A.S. Nazansky, *J. Chromatog.*, **116**, 99 (1976) **silica PC**
187 A.C. Ouano, *J. Macromol. Sci., Revs. Macromol. Chem.*, **C9**, 123 (1973) **PC calibration**
188 A.C. Ouano, W. Kaye, *J. Polymer Sci., Polymer Chem. Edition*, **12**, 1151 (1974) **viscometer**
189 A.C. Ouano, *J. Polymer Sci.*, Part A-1, **10**, 2169 (1972) **laser light-scatter**
190 A.T. Chamberlain and J.S. Marlow **EC polar solvents**
191 B.B. Wheals, C.G. Vaughan, M.J. Whitehouse, *J. Chromatog.*, **106**, 109 (1975) **fluorescence detection**
192 J.C. Steichen, *J. Chromatog.*, **104**, 39 (1975) **UV-fluorescence detector**
193 J.A.F. de Silva and N. Strojny, *Anal. Chem.*, **47**, 714 (1975) **fluorescamine**
194 M.A. West, *Internat. Lab.*, 45, (Jan/Feb 1975) **fluorimeter**
195 M. Weigele, S. De Bernado, J. Tengi, W. Leimgruber, *J. Amer. Chem. Soc.*, **94**, 5927 (1967) **fluorescamine**
196 T. S. Stevens, H. Small, *U. S. Pat.*, 3 966 596 (1976) **superficially sulphonated resin**
197 J.J. Kirkland, *Chromatographia*, **8**, 661 (1975)
198 J.Y. Chuang and J.P. Johnson, *Sep. Sci.*, **10**, 161 (1975) **sample size in GPC**
199 B.L. Karger, M. Martin, G. Guiochon, *Anal. Chem.*, **46**, 1640 (1975) **injection volume**
200 R.W. Kebler, *Optical Prop. of Synthetic Sapphire*, Linde Co, Union Carbide, NY **sapphire**
201 J.E. Purcell, H.D. Downs and L.S. Ettre, *Chromatographia*, **8**, 605 (1975) **injection in GC**
202 R.M. Cassidy and R.W. Frei, *Anal. Chem.*, **44**, 2250 (1972) **injector from ball valve**
203 H. Spaans, H. Terol, A. Onderwater, *J. Chrom. Sci.*, **14**, 246 (1976) **injector**
204 J.H. Knox, *Chem. Ind.* (London), 29 (1975) **microparticle columns**
205 P.A. Bristow and B.F. Williamson, unpublished work (1974)
206 *J. Chrom. Sci.*, back cover, (July 1976)
207 J.C.D. Brand and G. Eglington, *Applied Spectroscopy to Org. Chem.*, Oldbourne Press, London (1965), p. 107 **IR ϵ values**
208 J.K.F. Huber and J.A.R.J. Hulseman, *Anal. Chim. Acta.*, **38**, 304 (1967) **earliest ref.**
209 K. Callmer and D. Nilsson, *Chromatographia*, **6**, 517 (1973) **low-noise Varian**
210 J. Lehrfeld, *J. Chromatog.*, **120**, 141 (1976) **perbenzoyl derivatives**
211 ASTM D-20.70.04, Bibliography on Liquid Exclusion Chromatography, (1975), ASTM, Philadelphia
212 V.F. Gaylor, H.L. James, H.H. Weetall, *Anal. Chem.*, **48**(5), 44R (1976) **gel permeation review**
213 Y. Kato, S. Kido, M. Yamamoto, T. Hashimoto, *J. Appl. Polym. Sci.*, **19**, 629 (1974) **alumina in permeation chromatography**
214 R.E. Majors, *Internat. Lab.*, **7**(10), 11 (1975) **review LC packings**
215 Toya Soda Co., *HLC Report* 801, **1**, 36 (1974) (illustrated in ref. 214, p. 30) **HS permeation chromatography**
216 J.V. Dawkins and G. Taylor, *J. Poly. Sci.*, (B), **13**, 29 (1975) **alumina for PC**
217 H. Kalasz, J. Nagy and J. Knoll, *J. Chromatog.*, **107**, 35 (1975) **recycle**
218 R.E. Lovins, S.R. Ellis, G.D. Tolbert, C.R. McKinney, *Adv. Mass Spectrom.*, **6**, 475 (1974) **LC-MS**

219 A.C. Ouano, *J. Chromatog.*, **118**, 303 (1976) **laser light-scattering**
220
221 D.S. Horne, J.H. Knox and L. McLaren, *Sep. Sci.*, **1**, 531 (1966) **zone dispersion**
222 J.R. Cluff and S.J. Hawkes, *J. Chrom. Sci.*, **14**, 248 (1976) **zone dispersion review**
223 J.C. Giddings, *Anal. Chem.*, 37, 60 (1965) **theoretical separation limits**
224 K. Slais and M. Krejci, *J. Chromatog.*, **91**, 161 (1974) **solvent peaks**
225 H.J. Breter and R.K. Zahn, *Z. Anal. Chem.*, **279**, 151 (1976) **radioactivity as detector**
226 H. Oster and E. Ecker, *Chromatographia,* **3**, 220 (1970) **detector volume**
227 J.F.K. Huber, R. Van der Linden, E. Ecker, M. Oreans, *J. Chromatog.*, **83**, 267 (1973) **column switching**
228 M. Roth, *Anal. Chem.*, **43**, 880 (1971) **fluorescing reagent for amino acids**
229 J. Lankelma and H. Poppe, *J. Chromatog.*, in the press **glassy carbon coulometry detector**
230 D.R. Deans, *J. Chromatog.*, **18**, 477 (1965) **backflushing**
231 D.R. Deans, *Chromatographia,* **1**, 18 (1968) **heart-cutting**
232 S.T. Sie and N. van den Hoed, *J. Chrom. Sci.*, **7**, 257 (1969) **slurry packing silica**
233 P.B. Hamilton, *Anal. Chem.*, **30**, 915 (1958) **slurry packing ion-exchange**
234 C.D. Scott and N.E. Lee, *J. Chromatog.*, **42**, 263 (1969) **slurry packing ion-exchange**
235 R.E. Majors, *Anal. Chem.*, **44**, 1722 (1972) **slurry packing silica microparticles**
236 H.R. Linder, H.P. Keller, R.W. Frei, *J. Chrom. Sci.*, **14**, 234 (1976) **slurry reservoir**
237 J.H.Knox and A. Pryde, *J. Chromatog.*, **112**, 171 (1975) **bonded materials**
238 P.A. Bristow, *J. Chromatog.*, **122**, 277 (1976) **prep chromatograph**
239 M. Krejci and N. Pospisilova, *J.Chromatog.*, **73**, 105 (1972) **detector variance**
240 K.D. Mielenz, *Anal. Chem.*, **48**, 1093 (1976) **spectroscopy nomenclature**
241 E. Godbille and P. Devaux, *J. Chromatog.*, **122**, 317 (1976) **semi-prep columns**
242 W.H. McFadden, H.L. Schwartz, S. Evans, *J. Chromatog.*, **122**, 389 (1976) **LC–MS**
243 C.D. Scott, *Sep. Purif. Methods,* **3**, 263 (1974) **review HP ion-exchange in biochem**
244 R. Sargent and W. Rieman, *J. Org. Chem.*, **21**, 594 (1956) **salting out**
245 H.F. Walton, *Anal. Chem.*, **44**, 890 (1972) **ligand exchange**
246 P.J. Twitchett, A.E.P. Gorvin, A.C. Moffat, *J. Chromatog.*, **120**, 359 (1976) **pellicular and microparticle ion-exchange**
247 G.B. Cox, C.R. Luscombe, M.J. Slucutt, K. Sudgen, J.A. Upfield, *J. Chromatog.*, **117**, 269 (1976) **ion-exchangers on microparticle silica**
248 J.H. Knox and M. Saleem, *J. Chrom. Sci.*, **7**, 614 (1969) **theoretical predictions**
249 D.H. Saunders, R.A. Barford, P. Magidman, L.T. Olsezewski, H.L. Rothbart, *Anal. Chem.*, **46**, 834 (1974) **cation-ex on silica**
250 R.A. Hartwick and P.R. Brown, *J. Chromatog.*, **112**, 651 (1975) **microparticle anion-ex**
251 C.D. Scott, *Anal. Biochem.*, **24**, 292 (1968) **elutriation**
252 P.B. Hamilton, *Anal. Chem.*, **35**, 2055 (1963) **amino-acid analysis**
253 P.A. Asmus, C.E. Low and M. Novotny, *J. Chromatog.*, **119**, 25 (1976) **ion-ex on silica**
254 C. Horvath and S.R. Lipsky, *Anal. Chem.*, **41**, 1227 (1969) **ion-ex unpredictability**
255 H. Rückert and O. Samuelson, *Acta. Chem. Scand.*, **11**, 303 (1957) **non-aq. ion-ex.**
256 E. Päart and O. Samuelson, *J. Chromatog.*, **85**, 93 (1973) **IEC of sugars**
257 K.J. Bombaugh and P.W. Almquist, *Chromatographia,* **8**, 109 (1973) **thruput in prep**
258 L.R. Pickart and M.M. Thaler, *Prep. Biochem.*, **5**, 397 (1975) **prep LC of peptides**
259 A.R. Cooper, A.J. Hughes, J.F. Johnson, *Chromatographia,* **8**, 136 (1975) **prep PC**
260 J.J. De Stefano and J.J. Kirkland, *Anal. Chem.*, **47**, 1103A and 1193A (1975) **prep LC**
261 L.R. Snyder, *J. Chromatog.*, **92**, 223 (1974) **solubility and elution review**
262 J.C. Moore, *J. Polymer Sci.*, A, 2, 835 (1964) **classic GPC paper**
263 J. Porath and P. Flodin, *Nature,* **183**, 1657 (1959) **gel filtration discovery**
264 A.N. Martin, J. Swarbrick, A. Camarata, *Physical Pharmacy,* 2nd ed., Lea and Febiger, Philadelphia (1969) ISBN 8121 0163 4, p.516 **suspensions**
265 T.J.N. Webber and E.H. McKerrell, *J. Chromatog.*, **122**, 243 (1976) **slurry packing**
266 W.H. Morrison, M.F. Lou, P.B. Hamilton, *Anal. Biochem.*, **71**, 415 (1976) **sugar-borate IEC**
267 K. Unger and R. Kem, *J. Chromatog.*, **122**, 345 (1976) **silica for PC**
268 R.A.C. Gray, *Ph.D. Thesis,* Manchester (1976) **loading capacity**
269 F.E. Regnier and R. Noel, *J. Chrom. Sci.*, **14**, 316 (1976) **bonded silica for biological PC**
270 R.W. Frei and W. Santi, *Z. Anal. Chem.*, **277**, 303 (1975) **review derivatisation**
271 K. Unger, H. Fischer and E. Schadow, *Z. Phys. Chem., Neue Folge,* 99, 245 (1976) **critique of Hg porosimetry**
272 P.A. Asmus, G.E. Low, M. Novotny, *J. Chromatog.*, **123**, 109 (1976) **ion-ex on silica**

273 M. Maeda and A. Tsuji, *J. Chromatog.*, **120**, 449 (1976)
274 D.D. Bly, K.A. Boni, M.J.R. Cantow, J. Cazes, D.J. Harman, J.N. Little, E.D. Weir, *J. Polymer Sci. B*, **9**, 401 (1971) **definitions and equations in PC**
275 M. Nakajima, Y. Ozawa, T. Tanimura, Z. Tamura, *J. Chromatog.*, **123**, 129 (1976) **post-column derivatization detection**
276 D.L. Mays, R.J. van Apeldoorn, R.G. Lauback, *J. Chromatog.*, **120**, 93 (1976) **post-col reaction detection of kanamycin**
277 O. Samuelson and S. Johnson, *Anal. Chim. Acta*, **36**, 1 (1966) **Cr^{III}—dicarboxylic acid detection**
278 F. Nachtmann, H. Spitzy, R.W. Frei, *Anal. Chem.*,**48**,1576 (1976) **4-NBCl**
279 S. Siggia and R.A. Dishman, *Anal. Chem.*, **42**, 253 (1974) **pre-col deriv phenyl hydrazine**
280 A. Wolkoff and R.H. Larose, *J. Chromatog.*, **99**, 731 (1974) **oxidation of phenols**
281 K.W. Williams and R.C. Smith, *Prog. Med. Chem.*, ed. G.P. Ellis and G.B. West, **12**, 105 (1975), North-Holland, Oxford **review of affinity chromatography**
282 H. Amneus, D. Gabel, V. Kasche, *J. Chromatog.*, **120**, 39 (1976) **affinity**
283 C.R. Scholfield, *Anal. Chem.*, **47**, 1417 (1975) **prep with reverse-phase**
284 J.M. Bather and R.A.C. Gray, *J. Chromatog.*, **122**, 159 (1976) **silica surface**
285 M. Martin, C. Eon, G. Guichon, *J. Chromatog.*, **110**, 213 (1975) **pressure in LC**
286 V. Rehak and E. Smolkova, *Chromatographia*, **9**, 219 (1976) **review bonded phases**
287 J.B. Peri, *J. Phys. Chem.*, **69**, 211 and 220 (1965) **alumina surface**
288 P.A. Haas, M.H. Lloyd, W.D. Bond, J.P. McBride, *AEC* 31689 Report No ORNL-P-2159, *CA* 66: 22520; *Nucl. Sci. Abs.*, **20**(17) 3867 (1966) **sol gel**
289 J.M. Fletcher and C.J. Hardy, *Chem. Ind. (London)* 48 (1968) **sol gel**
290 P.F. Wace, Brit. Pat. 1 241 442 (1971) **spheres by sol-gel process**
291 W. Dress and J.H. Grimes, Brit. Pat. 1 175 834 (1969) **dextran added to sol**
292 P.F. Wace, Brit. Pat. 1 311 284 and 1 311 285 (1973) **engineering for sol-gel**
293 M.J. Fulwyler, *U.S. Pat.* 3 790 492 (1971) **microsphere production**
294 C. Hansch, *Accounts Chem. Res.*, 232 (1969)**biol. correlations with partition coefficients**
295 P.J. Taylor, M.S. Mirrlees, S. Moulton, C.T. Murphy, *J. Med. Chem.*, **19**, 615 (1976) **octanol—water**
296 D. Henry, J.H. Block, J.L. Anderson, *J. Med. Chem.*, **19**, 619 (1976) **partition coeff.**
297 J.M. McCall, *J. Med. Chem.*, **18**, 549 (1975) **partition coeff on ODS**
298 R.M. Carlson, R.E. Carlson, H.L. Kopperman, *J. Chromatog.*, **107**, 219 (1975)
299 R.M. Cassidy, *J. Chromatog.*, **117**, 71 (1976) **sulphur reverse-phase**
300 A. Nakae and G. Muto, *J. Chromatog.*, **120**, 47 (1976) **reverse-phase**
301 T. Hanai and K. Fujimura, *J. Chrom. Sci.*, **14**, 140 (1976) **org. and inorg. adsorbents**
302 R.B. Sleight, *J. Chromatog.*, **83**, 32 (1973) **reverse-phase on ODS**
303 K. Unger, J. Schick-Kalb, K.-F. Krebs, *J. Chromatog.*, **83**, 5 (1973) **silica production**
304 J.H. Knox, Lecture to the 11th International Symposium on Chromatography, Birmingham (1976) **ion-pairs**
305 J.H. Knox and G.R. Laird, *J. Chromatog.*, **122**, 17 (1976) **soap chromatography**
306 E.J. Kikta and E. Grushka, *Anal. Chem.*, **48**, 1098 (1976) **alkyl-bonded phases**
307 N.E. Hoffman and J.C. Liao, *Anal. Chem.*, **48**, 1104 (1976) **anilide derivatives**
308 E. Bayer, E. Grom, B. Kaltenegger, R. Uhmann, *Anal. Chem.*, **48**, 1107 (1976) **dansyl**
309 Y. Kasai, T. Tanimura, Z. Tamura, *Anal. Chem.*, **47**, 34 (1975) **derivative reagent**
310 R.G. Muuze and J.F.K. Huber, *J. Chrom. Sci.*, **12**, 779 (1974) **post-col. derivatives**
311 S. Katz and W.W. Pitt, *Anal. Letters*, **5**, 177 (1972)
312 A.W. Wolkoff and R.H. Larose, *J. Chromatog.*, **99**, 731 (1974) **Ce reagent**
313 R.W. Frei and J.F. Lawrence, *J. Chromatog.*, **83**, 321 (1973) **dansyl hydrazine**
314 S. Eksborg, P.-O. Lagerstrom, R. Modin, G. Schill, *J. Chromatog.*, **83**, 99 (1973) **ion-pair acids**
315 R.W. Frei, J.M. Huen, W. Santi, *J. Chromatog.*, in press **ion-pair**
316 S. Eksborg, G. Schill, *Anal. Chem.*, **45**, 2092 (1973) **ion-pair**
317 B.-A. Persson and B.L. Karger, *J. Chrom. Sci.*, **12**, 521 (1974) **ion-pair**
318 B. Fransson, K.G. Wahlund, I.M. Johansson, G. Schill, *J. Chromatog.*, **125**, 327 (1976)
319 J.C. Kraak and J.F.K. Huber, *J. Chromatog.*, **102**, 333 (1974) **ion-pair**
320 J.C. Meakin, personal communication
321 J.W. Higgins, *J. Chromatog.*, **121**, 329 (1976) **benzoyl derivative of sapogenin**
322 K.-F. Krebs and H. Heinz, (E. Merck, Darmstadt), *Ger. Pat.* 2 042 910 **silica sphere manufacture**
323 H. Halpaap, *J. Chromatog.*, **78**, 63 (1973) **structure of silica**
324 J.J. Kirkland, *J. Chromatog.*, **83**, 149 (1973) **Zorbax silica**
325 J.J. Kirkland, *Ger. Pat.* 2 317 455 **Zorbax patent**

326 W.A. Aue and S. Kapila, in E. Gruska ed., ref 10 pp. 13-25 **modification of silica**
327 A.V. Kiseler, *J. Chromatog.*, **49**, 84 (1970) **hydrothermal treatment**
328 *Biblio. on Controlled-pore glass chromatography*, Electronucleonics, Fairfield, N.J., USA (1974) **CPG biblio**
329 B.G. Linsen and J.C.P. Broekhoff in ref 24, chap. 1 **pore measurement**
330 S. Brunauer, P.H. Emmett, E. Teller, *J.A.C.S.*, **60**, 309 (1938) **surface area N_2**
331 J. Vermont, M. Deleuil, A.J. de Vries, C.L. Guillemin, *Anal. Chem.*, **47**, 1329 (1975) **Spherosil**
332 N.E. Fisher and A.Y. Mottlau, *Anal. Chem.*, **34**, 714 (1962) **pore volume**
333 H. Halpaap, *J. Chromatog.*, **78**, 77 (1973) **surface area measurement**
334 G.L. Hawk, J.A. Cameron, L.B. Dufault, *Prep. Biochem.*, **2**, 193 (1972) **polyethylene glycol deact. CPG**
335 F. Eisenbeiss, *Merck Kontackte* 17 (Feb 76) **surface area/pore size and selectivity**
336 I. Jane, *J. Chromatog.*, **111**, 227 (1975) **use of sedimentation**
337 R.P.W. Scott and P. Kucera, *J. Chrom. Sci.*, **13**, 337 (1975) **heating silica**
338 R.P.W. Scott and P. Kucera, *J. Chrom. Sci.*, **12**, 473 (1974) **survey of silicas**
339 J.J. Kirkland, *J. Chrom. Sci.*, **7**, 361 (1969) **Zipax CSP**
340 J.J. Kirkland, U.S. Pat 3 505 785 (1969) **Zipax patent**
341 J.J. Kirkland and C.H. Dilkes, *Anal. Chem.*, **45**, 1788 (1973) **ODN on Zipax**
342 J.D. Madeley and R.C. Richmond, *Z. anorg. allg. Chem.*, **389**, 92 (1972) **OH on silica**
343 E. Borello, G. Della Gatta, B. Fubini, C. Morterra, G. Venturello, *J. Catalysis*, **35**, 1 (1974) **alumina hydration**
344 G.J. Kennedy and J.H. Knox, *J. Chrom. Sci,*, **10**, 549 (1972) **comparison of packings**
345 J.J. Kirkland, *J. Chrom. Sci.*, **9**, 206 (1971) **ODS Permaphase**
346 J.H. Knox and G. Vasvari, *J. Chromatog.*, **83**, 181 (1973) **Permaphases**
347 C. Gonnet and J.L. Rocca, *J. Chromatog.*, **120**, 419 (1976) **ternary systems for LLC**
348 J.F.K. Huber, C.A.M. Meijers, J.A.M.J. Hulsman, *Anal. Chem.*, **44**, 111 (1972) **ternary**
349 C. Hesse and W. Hövermann, *Chromatographia*, **6**, 345 (1973) **ternary system for LLC**
350 H.N.M. Stewart and S.G. Perry, *J. Chromatog.*, **37**, 97 (1968) **LLC**
351 J.J. Kirkland, *Analyst*, **99**, 859 (1974) **trace analysis**
352 D.P. Wittmer, N.O. Nuessle, W.G. Haney, *Anal. Chem.*, **47**, 1422 (1975) **ion-pair**
353 K. Karch, I. Sebestian and I. Halasz, *J. Chromatog.*, **122**, 3 (1976) **bonded phases**
354 J.F.K. Huber, E.T. Alderlieste, H. Harren, H. Poppe, *Anal. Chem.*, **45**, 1337 (1973) **LLC**
355 N.A. Parris, *J. Chrom. Sci.*, **12**, 753 (1974)
356 B.B. Wheals, *J. Chromatog.*, **122**, 85 (1976) **ODS-silica and silica applications**
357 J. Shapiro and J.M. Kolthoff, *J.A.C.S.*, **72**, 776 (1950) **methyl red adsorption**
358 M. Lynn and A.M. Filbert in *E. Gruska*, ed., ref 10 **bonded phases on CPG**
359 R.K. Gilpin, D.J. Camillo, C.A. Janicki, *J. Chromatog.*, **121**, 13 (1976) **in situ bonding**
360 S.H. Chang, K.M. Gooding, F.E. Regnier, *J. Chromatog.*, **120**, 321 (1976) **oxirane bonding link**
361 C.H. Lochmuller and C.W. Amoss, *J. Chromatog.*, **108**, 85 (1975) **charge transfer phase**
362 B.B. Wheals, *J. Chromatog.*, **107**, 402 (1975) **vinyl silane**
363 H. Hemetsberger, W. Maasfeld, H. Ricken, *Chromatographia*, **9**, 303 (1976) **chain length of BP**
364 R.E. Majors and H.J. Hopper, *J. Chrom. Sci.*, **12**, 767 (1974) **BP (chain length)**
365 J.J. Kirkland, *Chromatographia*, **8**, 661 (1975) **ODS-Zorbax**
366 D.C. Locke, *J. Chrom. Sci.*, **12**, 433 (1974) **theory of BP separations**
367 R. Amos and S.G. Perry, *J. Chromatog.*, **83**, 245 (1973) **TLC—column**
368 W. Golkiewicz, *Chromatographia*, **9**, 113 (1976) **TLC—column comparison**
369 F. Geiss and H. Schlitt, *J. Chromatog.*, **82**, 5 (1976) **TLC—column summary**
370 B. Loev and M.M. Goodman, *Prog. Separ. and Purif.*, **3**, 73 (1970) CA 74: 68034x **dry-column**
371 C. Persiani, P. Cukor, K. French, *J. Chrom. Sci.*, **14**, 417 (1976) **glyceryl—CPG**
372 A.L. Vierk, *Z. Anorg. Chem.*, **261**, 283 (1950) **viscosity acetonitrile—water**
373 R.K. Gilpin and W.R. Sisco, *J. Chromatog.*, **124**, 257 (1976) **bonded-silane TLC**
374 D. Perrett, *J. Chromatog.*, **124**, 187 (1975) **aqueous gradient former**
375 R.P.W. Scott and P. Kucera, *J. Chrom. Sci.*, **11**, 83 (1973) **solvent—solute interactions**

376 N.K. Sabbagh and I.S. Fagerson, *J. Chromatog.*, **120**, 55 (1976) **glucose oligomer separ.**
377 R. Aiger, H. Spitzy, R.W. Frei, *J. Chrom. Sci.*, **14**, 381 (1976) **Ag–silica**
378 D.L. Ball, W.E. Harris, H.W. Habgood, *J. Gas Chromatog.*, **5**, 613 (1967) **peak area measurement**
379 L.C. Hansen and R.E. Sievers, *J. Chromatog.*, **99**, 123 (1974) **open-pore column**
380 S.R. Bakalyar and R.A. Henry, *J. Chromatog.*, **126**, 327 (1976) **quantitation errors**
381 H.K. Mangold and K.D. Mukherjee, *J. Chrom. Sci.*, **13**, 398 (1975) **TLC quantitation**
382 J.G. Kirchner, *J. Chrom. Sci.*, **13**, 558 (1975) **review of TLC**
383 R.P.W. Scott, *J. Chromatog.*, **122**, 35 (1975) **solvent–solute interactions**
384 O.G. Larionov and A.L. Myers, *Chem. Eng. Sci.*, **46**, 1025 (1971) **'phase'**
385 R.P.W. Scott and P. Kucera, *J. Chromatog.*, **125**, 251 (1976) **high plate no column**
387 J.N. Done, *J. Chromatog.*, **125**, 43 (1976) **sample load and HETP**
388 M. Martin, F. Verillon, C. Eon, G. Guiochon, *J. Chromatog.*, **125**, 17 (1976) **recycle**
389 J.J. Kirkland, *J. Chromatog.*, **125**, 231 (1976) **silica for PC**
390 J. Porath, *J. Macromol. Sci. Chem.*, A10 (1 & 2), 1 (1976) **affinity chromatog. review**
391 H. Colin and G. Guiochon, *J. Chromatog.*, **126**, 43 (1976) **carbon on silica**
392 R.E. Dessy, W.D. Reynolds, W.G. Nunn, C.A. Titus, G.F. Moler, *J. Chromatog.*, **126**, 347 (1976) **scanning UV detector**
393 I. Jane and J.F. Taylor, *J. Chromatog.*, **109**, 37 (1975) **fluorescence**
394 E. Fitzgerald, *Anal. Chem.*, **48**, 1734 (1976) **heptanesulphonate ion-pair**
395 J.C. Helmer, *Anal. Chem.*, **48**, 1741 (1976) **large piston pump and gradients**
396 A. McDowell and H.L. Pardue, *Anal. Chem.*, **48**, 1816 (1976) **fast scan UV detector**
397 D.M. Ordemann and H.F. Walton, *Anal. Chem.*, **48**, 1728 (1976) **aromatic hydrocarbons on ion-exchange resins**
398 J. C. Suatoni and R. E. Swab, *J. Chrom. Sci.*, **13**, 361 (1975) **backflush**

9.4 Trademarks

Trademarks are identified in the text by capitalizing the first letter. Apologies to any manufacturer whose trademark is not listed: not all companies replied to requests for trademark information.

trademark	company
Amcel	Celanese
Aminex	Biorad
AnalaR	Analar Standards
Arklone, Alkathene	ICI
Biogel	Biorad Labs
Bioglas	Biorad Labs
Biosil	Biorad Labs
Cajon	Crawford
Carpenter 20	Carpenter
CPG	Corning/Electronucleonics
Chelex	Biorad
Cheminert	Laboratory Data Control (LDC)
Chromotags	Regis Chemical
Corasil	Waters Associates
Coulter Counter	Coulter Electronics
Delrin	Du Pont
Dowex	Dow Chemical
Durrum	Durrum
Ferralium	Langley Alloys
Fluescin	Merck
Fluorinert	3M
Fluram	Roche Diagnostics
Fluon	ICI
Flutec	ISC Chemicals
Freon	Du Pont
Genklene	ICI
Gyrolok	Hoke
Hastelloy	Cabot
Ionex	Macherey-Nagel
Kel-F	3M
Kematal	Celanese
Langalloy	Langley Alloys
Licrosorb	Merck
Licrospher	Merck
Maranyl	ICI
Merkogel	Merck
Micropak	Varian
Miran	Wilkes
Monel	Henry Wiggin
Monel	International Nickel
Multiplex	Alpine
Nucleosil	Macherey-Nagel
Nupro	Crawford
Partisil	Whatman
Pellionex	Whatman
Pellosil	Whatman
Pellumina	Whatman
Penton	Hercules
Perisorb	Merck
Permaphase	Du Pont
Porasil	Whatman
Propathene	ICI
Refracto Monitor	LDC
Rigimesh	Trinity Micro
Sephadex	Pharmacia Fine Chemicals
Sepharose	Pharmacia Fine Chemicals
Sil-X	Perkin-Elmer
Sno-trik	Crawford
Spherisorb	AERE, Harwell
Spherosil	Rhone-Progil
Stellite	Cabot
Styragel	Waters
Swagelok	Crawford
Teflon	Du Pont
Tefzel	Du Pont
Trubore	Superior Tube
Vespel	Du Pont
Vilon	Du Pont
Vydac	Varian/Separations Group
Whitey	Crawford
Zipax	Du Pont
Zorbax	Du Pont
Zytel	Du Pont
17-4PH	Armco Steel
μ Styragel	Waters Associates

10 MATERIALS OF CONSTRUCTION

10.1 General Requirements

summary
- **stainless steels, fluorocarbons and ceramics suitable for most solvent combinations**
- **more exotic materials rarely worth considering except for critical components**

Corrosion resistance and strength to withstand high pressure are difficult requirements for materials with which to construct liquid chromatographs. It is the divers range of mixtures of solvents, acids, bases and salts that may be used that restricts the designers choice to a small number of materials, mainly some grades of stainless steel, fluorocarbon polymers and ceramics, which are freely available as components like tube and compression fittings. These compromise materials only limit the chromatographers solvent choice slightly. On balance high cost, poor availability in small sizes, and lack of all-round usefulness of other possible materials makes them unattractive except perhaps for some special applications or for critical components. This conclusion may change as other materials become more widely used in applied technology outside chromatography. At present it is better to try very hard to find an alternative chromatographic solution than stray from using these materials.

10.2 Stainless Steels

summary
- **316 and 321 most useful grades**

Stainless steels are not stainless. It is difficult to cause rusting in the most useful grades (usually with chromium about 18% giving a passive layer) but they are subject to other more insidious modes of corrosion.

The plain description 'stainless steel' is about as precise as the term 'plastic' and quite meaningless without type specification. (Instrumentation specification writers please note.)

Type selection is unusually awkward for metallurgists: it is the range of liquids and worse, the range of mixtures, inert individually but active in combination, that is the worst problem. Examples are halide impurities unionised and inactive in halocarbons, but ionised and active in mixtures with hydroxy solvents; and acids inactive in aqueous solution, but very corrosive in aqueous acetonitrile solution. Prediction is difficult: sulphuric-acetonitrile mixtures are rather corrosive but perchloric-acetonitrile mixtures comparatively inert. To be certain, tests with a particular material and solvent are essential, but in practice tell-tale signs suggest trouble before irrevocable damage to equipment is done. Experience also suggests that certain components and types of mix are to be avoided or used with caution.

Stainless steels come in a bewildering variety of types[56]. For LC applications several broad needs are identifiable apart from general corrosion resistance: machinability, ductility and work hardening, strength, high hardness (after heat treatment) for wear resistance and lower friction, weldability, and occasionally to be magnetic. No one grade meets all these needs. Cost at least is hardly a determining factor in choice of type for fabrication costs usually dominate.

Grades of special interest for LC are:

316 Good overall LC solvent corrosion resistance from added molybdenum; ductile where annealed; available in a very wide range of tube sizes and fittings; difficult to machine because of work hardening (though this is turned to advantage in making some compression fittings by rolling threads for higher strength); can be welded, though special low carbon grades (316L) are better for welding; non-magnetic.
321 Good overall corrosion resistance and especially low stress corrosion from added titanium; non-magnetic; similar to 316 but slightly cheaper; available in a wide range of tubes, but not fittings.

302 Basic 18/8 type (18% Cr, 8% Ni); less suitable for parts in direct contact with solvent systems.

304 Popular tube type especially as its ductility is ideal for making hypodermic needles which become stiff and easily sharpened by work hardening during drawing which also makes it slightly magnetic (can be reversed by annealing at about 1000°C, for example in a flame); less corrosion resistant than 316 or 321 but easier to machine especially in 'free cutting' grades containing added sulphur or selenium. Most useful for parts not normally in direct contact with solvent systems.

440C Nickel-free grade which can be made extremely hard, thus widely used for valve seats and balls. Unfortunately quite useless for LC because it is too easily corroded. Other martensitic grades like 410 S21 and 420 S45 can be hardened (from Vickers Hardness 200 when annealed to 540 after heat treatment) but are also easily attacked by solvent systems.

Precipitation Hardened Types

These combine a suitable composition and heat treatment to increase hardness and thus reduce wear (but at the price of lower corrosion resistance): thus widely used in piston pumps and some valve parts. In UK FV520B is the only readily available type but in the US a similar popular type is 17-4PH. Parts subject to high wear rates made in 316 or 321 give a shorter working life unless serious corrosion of the PH material takes place causing catastrophic seal wear and failure. Hard chrome plating on top of 17-4PH is used with success by Haskel Engineering Inc. who make pneumatic amplifier pumps for Dupont Instruments Division chromatographs.

Other Metals and Alloys

A few of the more exotic metals and 'superalloys' developed to combat corrosion in process chemical plant may occasionally be useful for LC especially for a small but critical part. Pump pistons are a typical problem area where these materials can be valuable.

Titanium and tantalum, and gold and other precious metals and their alloys are attractive for corrosion resistance but are not available in small size tubes and compression fittings except at high cost and to special order. The softness of gold and tantalum recommend them as gasket materials but in practice PTFE is a better

Related specifications for wrought chromium-nickel stainless steels

American	British*		French	German	Italian	Swedish	Japanese
AISI	Old En No.	BS 970 BS 1449	AFNOR	VD Eh designation Reference number	UNI	SIS	JIS
302	58A	302 S 25	Z 12 CN 18–10	X 12 CrNi 18 8 1.4300	X 15 CN 1808	2330	SUS 40
303	58M	–	Z 10 CNF 18–10	X 12 CrNiS 18 8 1.4305	X 15 CNF 1808	2346	–
304	58E	304 S 15 304 S 16	Z 6 CN 18–10	X 5 CrNi 18 9 1.4301	X 8 CN 1910	2332	SUS 27
304L	–	304 S 12	Z 3 CN 18–10	X 2 CrNi 18 9 1.4306	X 3 CN 1911	–	SUS 28
316	58J	316 S 16 320 S 17	Z 6 CND 18–12	X 5 CrNiMo 18 12 1.4436	X 8 CND 1712	2343	SUS 32
316L	–	316 S 12	Z 3 CND 18–12	X 2 CrNiMo 18 10 1.4404	–	–	SUS 33
321	58B 58C	321 S 12	Z 10 CNT 18–10	X 10 CrNiTi 18 9 1.4541	X 8 CNT 1810	2337	SUS 29

*The revised versions of BS 970 and BS 1449 include a more comprehensive code of designations which replaces the system of designating steels by En numbers. The first three figures of the new code are the group number and, wherever possible, correspond with the equivalent AISI number, if any. The following letter only indicates the broad class in which the steel is included, e.g. S for stainless steel, and has no relation to any letter following an AISI number. The last two figures designate the particular alloy composition within the basic group, thus allowing for variants such as low-carbon or free-machining steels.

choice unless high temperatures (above those used in LC) or wide changes in temperature are envisaged. Some cobalt-palladium alloys are magnetic and yet highly corrosion resistant, and so might be used in solenoid valves.

Monel, a nickel/copper alloy might be considered for applications where eluents containing concentrated chloride ion must be used and is available in a reasonable range of tube and compression fittings. Hastelloy C provides outstanding chemical resistance from a nickel/chromium/molybdenum alloy but poses severe machining problems caused by work hardening for the very small parts required for LC. Injection valves have been made with this material for sliding faces. High nickel content in a stainless steel, typically Langalloy 20V (20% Cr, 29% Ni), provides additional corrosion resistance against inorganic and organic acids. High chromium content as used in Feralium alloys also gives high resistance especially to chloride ion attack. It is also very hard and has good anti-friction properties.

Machining Stainless Steels[57]

The toughness of these alloys requires powerful and rigid machines. Work hardening occurs during machining and very sharp tools with suitable cutting angles and correct lubricants are needed to avoid problems. If the tool is allowed to dwell without cutting a hard zone builds up which is difficult to remove. It may be necessary to cut underneath the hardened zone. Experienced machinists will avoid these pitfalls but the material characteristics still impose limits on what can be achieved. In LC this is usually a limit to the length of small diameter holes that can be drilled (without resorting to spark erosion or laser drilling). It is usually best to avoid drilling difficult holes by inserting small bore tubing into a larger hole which can be drilled easily and reliably. Joints can be sealed by welding or brazing; sometimes a push fit will suffice where the high pressure from the pump is not applied to the seal. Alternatively difficult holes can be drilled in ptfe or ptcfe inserts and a seal made by clamping stainless parts against the plastic insert. Small holes and small volumes are essential to good design of valves, injectors, column end fittings and detector flowcells.

Welding

Though widely used in process chemical plant construction and ideal for large containers like reservoirs, welding of many LC componets is impractical because they are too small, or more commonly because a very small tube is to be joined into a large block of metal. Surface oxidation is an undesirable side effect. Electron beam welding can sometimes provide controlled and localised heating (and the vacuum environment prevents oxidation) but is expensive and difficult to feed filler metal if this is needed. Low-carbon stainless steels, for example 316L (BS 316 S12) grade, is desirable for deep welds.

Brazing

Brazing employs a jointing metal of much lower melting point than the main metal components. Normal flame heating methods are not suitable because of surface oxidation (which spoils the action of compression fitting threads, for example).

Commonly used braze materials using copper and silver are unsuitable because of corrosion by LC solvents, and they may contaminate the column with dissolved metals. Higher melting point alloys with high excellent corrosion resistance based on nickel, chromium and gold can only be used effectively in vacuum induction furnaces. Braze powder and flux paste can be applied very accurately and its flow controlled well. Since a furnace can hold many components this is an economical process for production of reasonable quantities at a time, and this process is used by several manufacturers, especially in attaching 1/16 in tube to valve bodies. BS 1723 : 1963 specifies the process and BS 1845 : 1966 suitable alloys. Au-Ni brazes offer very high corrosion resistance for joining 300 series stainless steels but is expensive and Ni-3 (BS) or B-Ni-2 (ASTM) is perhaps a better price compromise for marginally less corrosion resistance. Temperatures above 1000°C should be used with caution as they may cause annealing of work-hardened threads thereby reducing their strength in subsequent use at ambient temperatures.

Alloys for Vacuum-Induction Brazing (from BS 1845 : 1966)

Type	Composition	melting point °C
Ni 2	Cr 12, P 10, Ni balance	890
Ni 3	Cr 7 Si 4, B 3 Fe 3, Ni balance	955-1000
Au 6	Cu 25, Au 75	950-990

Springs

Springs pose special problems for the designer because suitable grades cannot be heat hardened as is normal when making springs and usually one must rely on work hardening. The temptation to use a better but more easily corroded spring should be resisted. Work-hardened 302 grades and 17-4PH have been satisfactory.

10.3 Plastics

10.3.1 Fluorocarbons

PTFE

summary
- **ptfe an important but tricky material to use**
- **seals must accommodate creep or flow and thermal expansion**

Polytetrafluoroethylene (PTFE)[61] is an important material for modern LC: but it is also a rather tricky material to use well. In the past many components have been unreliable because critical design details have been ignored. PTFE is famous for its inertness to all conceivable LC chemicals. It is also notorious for slow creep under load so that seals begin tight, and relax and leak. Less well-known is its abrupt change in size, about 1% linearly, at 19°C, a temperature that lies within the variation of ambient temperature in many laboratories. Both properties mean that dimensional stability cannot be assumed. Designs using PTFE must accommodate any dimension changes induced by load and temperature. There are many ways of achieving this but a few examples may be helpful both in assessing and understanding commercial designs and in building special equipment. Gasket design using PTFE (and PTCFE) is discussed by Anderson et al[60]. At stresses from 6-30 MN/m^2 PTFE behaves as a highly viscous fluid and flows to seal even quite a rough surface, and after flow the slight permeability of the material is decreased. This pressure must be maintained to maintain the seal and further flow contained by design of small clearance around the gasket, or further accommodated by some spring. Various spring systems are used, from the natural spring of polypropylene threads in the Cheminert type to disc springs which exert a carefully controlled and calculable loading. Many flanges also have sufficient intrinsic springs as they deflect when clamping bolts are tightened, so many seals work without conscious design.

O-rings in PTFE are rarely satisfactory when they replace elastomer o-rings directly. Internal springs of metal coils (as used in several LC designs, for example the Milton-Roy minipumps); internal elastomer o-rings with a sealed on PTFE outer, or even compressed gas (used in some Precision Sampling Corp. syringes) are all elegant solutions.

Valves that involve sliding faces, either both of PTFE or one harder face in metal or ceramic against PTFE, need a constant force of around 10 MN/m^2. An effective method used in several commercial designs uses disc springs, which are described and specified in DIN 2093[54, 55]. These are dished washers that generate a well defined force as they are flattened. They can be stacked to multiply force if all facing the same way, or to give higher deflection but constant force if alternating face to face. (Thus it is important to note how the springs are arranged on a commercial design as it is being disassembled, and reassemble correctly.) As a very rough rule of thumb each spring will load an area of PTFE not greater than that of the hole in the spring. (Filled PTFE will require higher loading.)

Examples of Disc Springs to DIN 2093 : 1967 all dimensions in mm

outside diameter	inside diameter	free height	thickness	deflection for force mm	N
8	4.2	0.6	0.4	0.15	200
12.5	6.2	1.0	0.7	0.19	650
16	8.2	1.25	0.9	0.26	1000
22.5	11.2	1.75	1.25	0.37	2200

PTFE is a high molecular weight, highly crystalline polymer with a melting point well over 300°C. The product of polymerisation is a fine powder 10-100 μm or a dispersion (0.25 μm) in a liquid. Because of its very high molecular weight it will not flow fast enough even above its melting point and so processing methods used for most other plastics are not applicable. The particles are first pressed together at room temperature giving a preform strong enough to be handled. This is then strengthened by heating to 380°C when sintering takes place and particles coalesce. Moulding is used for preparation of bar and components like o-rings and ferrules and seals. Because sintering causes some shrinkage very precise control of dimensions is impossible. Strains set up during moulding may make annealing of large blanks or roughed-out components desirable so that these strains are relieved before final machining.

Tube and small diameter rod are made by continuous moulding where the preform is immediately pressed through a long heated die where sintering is completed. Slight variations in the sintering efficiency cause PTFE tube to fail under high pressure by becoming slightly porous before mechanical collapse. Precise control of outside and inside diameter of tube is difficult and quite large variations are common in commercial tube used in LC. Tube that is slightly too large can be reduced to fit compression fittings by stretching the tube holding the two ends with pliers until the required diameter is obtained (not under tension of course) and the damaged ends discarded. The concomitant internal diameter reduction is normally unimportant. This method is easier and more effective than drawing through a conventional wire drawing die.

Tubing can be gripped by compression fittings, but these tend to cause thin wall tube to collapse and the internal diameter can be easily reduced to zero, a not uncommon cause of blockages. Flanging the tube end avoids this problem and is recommended for sealing to flat faces. The Cheminert range of fittings first exploited this idea most successfully and has been widely imitated. The elasticity of the actual moulded polypropylene fitting provides the spring force to seal and accommodate any creep of the PTFE flange. The internal volume and unswept volume are all commendably low, somewhat better than a standard (unbored) 1/16–1/16 in compression fitting, though bored fittings and the inverted, GC or low volume stainless fittings can also have good flow patterns. To form the flange a tool with a suitable pointed shape needs to be heated to 350-400°C. Most makers of the fittings supply tools for this purpose. Curie point temperature controlled soldering iron bits, machined to shape, heat up more quickly and give good temperature control at low cost. A Weller No 7, nominally 370°C, is suitable. Other PTFE components can also be formed this way. Rapid cooling, for example against a cold metal plate or former is essential or the flange will relax.

Larger diameter tube can be formed by careful glass blowing methods in a bunsen flame burner followed by quench cooling under cold water. Heating PTFE above 400°C causes degradation giving off toxic fumes so these operations should be carried out in a fume cupboard. Hazards of PTFE at high temperatures have been exaggerated[64]. Ventilation of heating operations and preventing dust or turnings getting mixed with tobacco are the only precautions LC users need take.

Filled PTFE

summary • **much improved wear resistance and less creep**

The incorporation of fillers improves wear resistance approaching a thousandfold, reduces creep, increases thermal conductivity and reduces thermal expansion[62]. Commonly used fillers are glass fibres, carbon or graphite, metals like bronze, and metal oxides. Particles sizes are usually from 5-50 μm. Up to 40% filler by weight may be added but about 25% is more common[65].

Valve seats, and sliding-plate valve faces, where resistance to creep and deformation is more important than wear resistance, are often made from 25% glass fibre grades. These may have metal oxides added to colour them primarily as a distinguishing feature; for example ICI Fluon grade VX1 is blue, whereas Rulon, a similar material, is red.

For piston seals on reciprocating pumps, wear resistance is more important and bronze and/or graphite filled formulations are more common, for example Fluon grade VX2 (bronze/graphite) or VP25 (carbon). The many graphite filled formulations offer excellent solvent and wear resistance.

The effect of solvent lubricating properties makes it difficult to predict the best filler for general LC applications, but any filler is likely to give at least a tenfold improvement over unfilled material. PTFE seal design is not straightforward but our understanding is increasing though operation under the actual conditions of use is the final test. The seal of PTFE is made on a different principle to that of elastomers in that the material is truly self-lubricating. Some material is transferred from the seal to the metal surface to fill the valleys in surface roughness so that the final rubbing surfaces are largely PTFE against itself. Surface finishes of 0.2–0.5 μm (8–16 μin) to BS 1134 CLA (centre line average) produced by grinding or lapping (not fine turning) are recommended. There is some evidence that smoother finishes give a worse seal.

Porous PTFE

This useful material can be produced by sintering powder at about 380°C under a controlled pressure of 15–30 MN/m^2. By varying the granule size and the sintering conditions pore sizes from 10 to 300 μm with pore volumes of 5–35% can be produced. It is difficult to control the resultant porosity and values given are highly nominal though reproducible if produced by the same processor on the same equipment. Two processors whose material is used for chromatography are Fluoro-Plastics Inc, Philadelphia, Pa, USA, and Vycrapor Ltd, Irvine Industrial Estate, Ayrshire, Scotland.

Coining

Although PTFE cannot be melted, heating to 310-320°C reduces the force required to cause flow and parts like valve seats can be coined using pressure in the range 15–40 MN/m^2. A complex shape with excellent surface finish can be produced. Even at room temperature its flow properties can be exploited to form small parts which are difficult to machine. Thus compression fittings can be used to mould parts to fill their unwanted internal voids using wire, rod or tube as a former where flow passages are required.

Shrink fitting

The high temperature coefficient of PTFE (and other plastics) can be exploited by shrink fitting components into metal parts. Both parts are cooled to –78°C in a solid CO_2/methanol bath or –196°C in liquid nitrogen and assembled. As the plastic expands more than the metal it will be gripped tightly on warming. Parts can be machined to 1–2% oversize, for example 0.01 in on a 1 in diameter component. Conversely internal seals can often be removed by cooling in a refrigerator or freezer. Even cold water will often take the material through the 19°C transition reducing size by nearly 1%.

Physical Properties of PTFE (pure and filled)

	Units	Unfilled	VG25	VP25
density	g/cm^3	2.17	2.25	2.1
filler	% w/w	–	25% glass	25% carbon
max. tensile strength	MN/m^2	20–35	12–20	10–15
Elongation at break	final length per unit initial length	3.5–5	3–4	2
Hardness	Shore D	50–55	55–70	60–65
Linear thermal expansion coefficient below 20°	K^{-1}	10×10^{-5}	5×10^{-5}	7×10^{-5}
thermal conductivity	W/(mK)	0.25	0.4	0.6

Polychlorotrifluoroethylene

summary • **PCTFE more elastic and stronger than PTFE but more expensive**

Kel-F[67] is a commercial thermoplastic resin formed by polymerisation of chlorotrifluoroethylene. Its chemical resistance is almost as high as PTFE, but its strength and elasticity, and cost, much higher. Apart from price, it is ideal where the mechanical strength of PTFE is too low. Only at higher temperatures (>100°C) do carbon tetrachloride and diethyl ether dissolve in it and cause some swelling and softening. For practical purposes it can be considered completely resistant for LC applications.

Unlike PTFE it can be injection moulded at 250-280°C with suitable equipment. Alas this process is only economic for large numbers of parts. It melts at 249° and is thermally stable to 280°C. Above 340°C it is rapidly degraded. Though it has a high thermal expansion coefficient like all plastics it has no room temperature phase transition and can be machined to a higher accuracy than PTFE. Flat and smooth faces can be produced by polishing or lapping and by heating and cooling under pressure against a smooth mould. Flanging of tube with techniques like PTFE but at lower temperatures is also possible. (A Weller No 6 Curie point tip nominally 310°C is most suitable.)

Kel-F also flows under stress like PTFE but at about 40 MN/m^2, double that for PTFE. Similar loading techniques must be used to avoid leaks[60]. At low stresses it is much more elastic than PTFE and so easier to design with.

Other Fluorocarbons

Copolymers of tetrafluoroethylene and hexafluoropropylene, fluorinated ethylene propylene, and others are becoming available. At present they do not appear to have any especially desirable features for LC equipment that PTFE and PCTFE lack.

10.3.2 Other Plastics

Polypropylene

This material is best used for hot moulded parts. Its solvent resistance is good but not equal to PTFE; its mechanical strength less than other plastics (which can be exploited in that replaceable polypropylene threaded fittings strip before the irreplaceable body). Weldability can be useful but welds are much less resistant than solid material and is only fair under combined stress and corrosion. Welded solvent reservoirs have proved unsatisfactory.

Glass fibre reinforced material has better strength but is unpopular in the workshop as the glass blunts tools rapidly. Diamond or tungsten carbide tipped tools have been used successfully. Thus this attractive material is less used in LC than in higher volume production applications. A suitable grade is Propathene HW 70GR, and there are many others.

Polyacetal, Nylon and Polyester plastics

These materials have much better mechanical and machining properties than PTFE but much lower chemical resistance (and lower cost). Both properties are further improved by glass fibre and/or carbon fillers. Thus they are widely used for the body part of valves and other components used in LC, while all parts exposed to liquid are usually in more resistant but expensive materials. Small seepages usually evaporate before too much damage is done, although when using concentration sample solutions, or salt buffered solvent the residue after evaporation can have disastrous results on sliding surfaces. Flushing the 'outside' surfaces with water is used effectively in some commercial valves to prevent crystallisation of traces of buffer. Despite the promises manufacturers give, all sliding seals leak: it is essential for the device to tolerate this without further damage.

Typical materials are Delrin (acetal homopolymer); Kematal (acetal copolymer); Deroton (polyester) (TAR 15, unfilled[70]; or TGR 80, 30% glass-filled[69]); Nylon 66 (Maranyl A100, unfilled[71]; AD390, glass-filled; A198 glass-fibre and graphite filled[72]) and there are many other similar materials made with other trade marks. For LC applications the differences between different materials are usually quite small[73].

10.4 Ceramics

summary
- **alumina (sapphire) pump parts corrosion and wear resistant**
- **glass weak and fragile for making columns**
- **alumina stronger than silica**

Ceramics are used for their high hardness and wear resistance in pump and sample injection valves, and pump pistons. Synthetic sapphire (alumina) is used in several instruments: it can be machined to very high accuracy on special equipment with diamond-based cutting tools. (Wear of metal components often takes place because the wiping action of seals removes passive surface layers thereby continually exposing new surface to corrosion: material suitable for static components like pipework may be unsatisfactory for moving seals.)

Glass has been used for columns where its smooth bore is believed to reduce irregular packing against an irregular wall surface. The main disadvantage is its low tensile strength and fragility under impact, so pressure limits are low, especially for large diameters. Connections to glass are problematic and not all commercial designs are satisfactory. Glass-lined stainless steel tube combines the virtues of the two materials but diameters available are rather small to be very useful.

Silica and alumina are used as optical windows in many detectors: both have excellent transparency down to 190 nm. Alumina also has an interesting transparency out to ~7 μm which includes a number of intense absorption bands which might be used for infrared detection; it is also compatible with all LC solvents. Alumina[200] is gaining popularity as a window material (and for mechanical components) because it is much stronger than silica and so is less likely to break under the stress of forming a seal or from high liquid pressure. However it is still more expensive than silica.

Forming seals between ceramics and other components is a persistent problem especially as PTFE is the usual choice of gasket material. As noted in section 10.3 a spring loading system is essential. Disc springs[54,55] have been used to load flowcell windows against a gasket. The Cheminert pattern of fittings also seals well against smooth ceramic surfaces.

Silica components can be stuck together to form 'fused quartz' cells. This process allows more complex components to be assembled than could be made by machining, blowing or moulding techniques. The parts are cut to size and faces to be joined ground optically flat. When assembled the parts adhere and a very strong joint results from heating to ~400°C, well below the softening point. This method is being extended to alumina.

10.5 Tubing

summary
- **seamless 316 stainless steel tube to ASTM A269 suitable for pipework**
- **cold drawn tube has better internal finish**
- **smooth precision bore expensive but reduces wall effects in long narrow columns**

Tubing is conventionally sized from its outside diameter; pipe is sized by its nominal bore. The smallest pipe size is 1/8 in bore, too large for plumbing in LC. A controlled outside diameter is needed to use compression fittings with tube; it is normal practice to tap the bore, or die the outside of pipe to make connections, or to weld. Outside diameter or od will be referred to simply as diameter elsewhere in this book; the internal diameter will be called the bore. Wall thickness is also often used together with the outside diameter to specify tubing, and is often quoted in terms of a sheet metal or wire gauge number.

Standard Wire Gauge SWG
Conversion table

SWG	mil or thou	mm
10	125	3.18
14	80	2.03
16	64	1.63
18	48	1.22
19	40	1.02
20	36	0.91
22	28	0.71
24	22	0.56
26	18	0.46
28	14	0.36
33	10	0.254

Tube can be drawn from a solid block of metal, when it is called seamless. This type is most suitable for LC because it has a smooth bore. Tube that is made by welding sheet to form a tube and then drawn down to small diameter tube is also suitable. Pipe that is several inches in diameter is usually welded. The weld leaves a rough area which may need to be machined off if a smooth bore or outside diameter surface is needed.

Cold drawn tube generally has a better surface finish on both the outside and the inside. BS 1449:Pt 2:1975 describes and helps specify stainless steels and their surface finish. Bright annealed or finish 2C is most suitable for making columns. Hot drawn tube, annealed and descaled (finish 1) is suitable for general pipework but the coarser internal finish is less suitable for making columns. Descaled tube is acid treated to remove surface oxidation during heat treatment to anneal tube. Tube is always supplied in the annealed condition.

Most tubing is supplied to ASTM A269 (1971); BS 3605:1973 and DIN 1629 Blad 3 on tubes and DIN 2448 on seamless tubes are also relevant. Very small diameter tube ($<$1/32 in) is called hypodermic tube, to which the above specifications do not apply. BS 3552:1962 gives some information on this type. Hypodermic tube may not be annealed and so be much stiffer than seamless tube.

Precision bore tube is also available though much more expensive and often to special order only in a restricted range of sizes. Its smooth internal bore (<1 µm roughness compared to ≈ 10 µm for cold drawn seamless tube) reduces wall effects in non-infinite diameter columns. It is widely used for long narrow columns used with quite large pellicular packings. Polishing or drilling is said to reduce wall effects. Different packing methods, for example dry or slurry methods techniques, will alter the effect of surface roughness so it is difficult to make general statements on the importance of bore finish.

Inch size tubes are available in a wider range of sizes than metric tubes, especially in the number of wall thicknesses stocked.

Inch tube sizes in seamless stainless steel

diameter in	mm	wall in	mm	bore in	mm	clearance drill size	area mm^2	length containing 1 µl
1/32	0.79	0.004	0.10	0.023	0.58	No 67 or	0.26	2.9 mm
		0.006	0.15	0.019	0.48	0.8 mm	0.18	4.4
0.031		0.008	0.20	0.015	0.38		0.11	7.1
0.040	1.02	0.008	0.20	0.024	0.61	No 58 or 1.1 mn	0.29	6.3
1/16	1.59	0.010	0.25	0.043	1.09	No 52 or 51	0.93	0.84
0.063		0.016	0.41	0.031	0.79	1.6 mm	0.49	1.60
		0.022	0.56	0.018	0.46		0.16	4.7
1/8	3.18	0.006	0.15	0.113	2.87	No 30 or 29	6.4	0.12
0.125		0.028	0.56	0.069	1.75	3.2 mm	2.4	0.33
		0.049	1.24	0.027	0.69		0.37	2.1 mm
								210 µm
1/4	6.35	0.028	0.71	0.194	4.93	letter F	19	41 µm
0.250		0.036	0.91	0.178	4.52	or 6.5 mm	16	49
		0.064	1.63	0.122	3.1		7.5	104
		0.080	2.03	0.090	2.29		4.1	191
		0.094	2.39	0.062	1.57		1.9	404
3/8	9.53	0.028	0.71	0.319	8.11	letter V	51	15.1
0.375		0.048	1.22	0.279	7.09	or 9.6 mm	40	20
		0.080	2.03	0.215	5.46		23	34
		0.125	3.18	0.125	3.18		8	100
1/2	12.7	0.028	0.71	0.444	11.28	12.8 mm or	100	8
0.500		0.048	1.22	0.404	10.26	bore to	83	10
		0.094	2.39	0.312	7.92	0.505 in	49	16
1	25.4	0.048	1.22	0.904	22.96	bore to	410	2
		0.064	1.63	0.872	22.15	1.005 in	390	2
		0.080	2.03	0.840	21.34	or 25.5 mm	360	2
		0.125	3.18	0.750	19.05		285	3
PTFE tube sizes								
1/16	1.63	0.016	0.41	0.032	0.81		0.52	1.94 mm
0.063		0.026	0.66	0.012	0.30		0.070	14.3
		0.029	0.74	0.006	0.152		0.018	55.5
1/8	3.17	0.032	0.81	0.064	1.63		2.08	0.53
0.125								

Metric tube sizes in seamless stainless steel

diameter	wall thickness	bore	clearance drill size	bore area	length containing 1 μl
6	1	4	6.1 or	12.6	0.08
	2	2	letter B	3.1	0.32
10	1	8	10.1 or	50.2	0.02
	2	6	letter X	28.3	0.035
12	1	10	12.1 or	78.5	0.01
	2	8	31/64 in	50.2	0.02

All dimensions in mm

Hypodermic Tubing

gauge	diameter thou	mm	bore thou	mm	wall thou	mm
26	18.5	0.46	9	0.24	4	0.11
26	18.5	0.46	4.5	0.14	7	0.18
25	20	0.51	8	0.2	6	0.13
23	25	0.64	8	0.2	8.5	0.16
22	28	0.71	6	0.15	11	0.28
22	28	0.71	16	0.4	6	0.16

Bursting

Working Pressure of Stainless Steel Tubing

Working pressures are calculated from experimentally determined bursting pressures by applying an arbitrary safety factor. This factor varies with the conditions of use, whether any corrosion is likely, and whether liquid or gas pressures are involved. The table below is drawn from ASA 31.3 (1959) table 302.2.1A and paragraph 304.1.2. and applies to cold drawn seamless tube 304 or 306 grade in the fully annealed condition with an ultimate strength of 18 750 psi, over the range 0 to 40°C. The bursting pressure is divided by a safety factor of 4 to give a safe working pressure. For LC applications which are usually liquid pressure and not subject to corrosion these ratings are conservative especially as the consequences of a tube burst with many types of pumps are unlikely to lead to hazard.

diameter inch	wall thickness thou 20	28	35	49	65	83	120
1/8	6800	10 000	12600				
1/4		4 300	5900	8640	11700		
3/8			3780	5470	7500		
1/2			2700	4000	5400	7180	
1			1350	1890	2570	3340	5000

pressures shown are in psi

10.6 Pipe Fittings

summary
- **following assembly instructions improves reliability**
- **nut and ferrule(s) must be same make**
- **tube will burst before fitting yields**

Nearly all chromatographs now use compression fittings to connect tubing. There are several manufacturers of these fittings: compatibility is a problem and an international standard badly needed. Four manufacturers fittings can be interchanged in as much as the thread sizes are the same, but the ferrule–nut interaction is particularly important for a reliable seal so the nut and ferrule must be the same make. It is plausible to assume that the highest reliability will be achieved by using only components from one manufacturer but despite strident claims and counterclaims by the manufacturers there is no reliable statistical evidence.

A compression fitting works by deforming a ferrule (or ferrules) so that it bites into the tube (usually in two places to resist vibration) and grips it, preventing the tube being blown out and forming a seal. If the wall is thin the tube bore may be deformed as well. The tube should not turn as the nut is tightened to force the ferrule into the tube, which must be soft enough (usually annealed) so that the ferrule can cut in. To reduce the torque required, and to avoid galling of the thread (a sort of friction weld) which is a serious problem with stainless steel threads, the nut is lubricated with silver plating or a molybdenum disulphide coating baked on. Correctly assembled all types will hold and seal up to the bursting pressure of the tube.

All fittings have a 'standard depth' for each pipe size so that once a ferrule is swaged on, it can be reconnected into any other fitting. Any machining may alter this depth and thus the distance from tube end to ferrule, so that the combination is no longer interchangeable. Reversed, chromatography, or low dead volume fittings are used where dead volumes are critical. Drilled out fittings where two tubes butt directly together achieve an even smaller volume with no change in bore.

To assemble a compression fitting

The tube should be cut or sawn reasonably square and filed or scraped free of burrs and swarf. The tube should pass easily through the ferrule(s) and be seated in the body without binding. The nut should first be tightened by hand, and then with spanners 1— 1 1/4 turns (3/4 turn for 1/16 and 1/8 tube sizes). If the fitting is undone and reconnected it should be necessary to tighten only 1/8 turn beyond hand tightness.

A common problem is that the tube sticks in the fitting after the nut is completely undone. This is often caused by failing to remove all burrs on the outside. As the nut is tightened the tube is pressed hard into the body which can jam it in: thin wall tube tends to buckle and jam for this reason. This force can be exploited to grip a mesh disc in a column end fitting. The other common problem comes from overtightening which pulls first the body and eventually the nut threads. It pays to follow the manufacturers instruction to mark nut and body so that the fitting is only tightened the specified amount. Taking this simple step and keeping the sealing surface clean will improve reliability—but no type is perfect and none last for ever.

PTFE Ferrules

The axial force pushing the tube against the seating shoulder of fittings can cause glass tube to crack. This can be avoided by keeping the tube a half millimetre short of the shoulder as the nut is tightened.

Compression Fitting Threads

The life of pulled threads can be extended somewhat by dieing the thread until a new nut will go on by hand. Damaged nuts propagate distorted threads and must be discarded immediately.

In general it is best to use actual compression fittings to construct special assemblies which need to connect to pipes even if it means brazing or welding fittings to other machined parts. But sometimes it is more convenient to fashion a part threaded to accept a compression fitting nut. Approximate thread sizes are listed below. The pitch and diameter are correct but the thread form is slightly different. It is difficult to machine a special part to exactly the thread form of a compression fitting (many makes are rolled rather than cut). Metal ferrules tend to give less reliable joints but PTFE ferrules deform to accommodate slight differences (but will not work at as high pressures as stainless steel ferrules).

Useful thread sizes

type	diameter in	mm	threads/in	pitch mm	tapping drill	
1/4 UNF	.250	6.35	28	0.91	7/32 in or	5.5 mm
1/16 NPT	.310	7.87	27	0.94	I	7
1/8 BSP	.383	9.72	28	0.91	5/16	8
1/8 NPT	.405	10.29	27	0.94		
1/4 BSP	.518	13.16	19	1.34	29/64 in or	11.8 mm
1/4 NPT	.540	13.72	18	1.41		
3/8 BSP	.656	16.66	19	1.34	19/32	15.3
3/8 NPT	.675	17.15	18	1.41		
1/2 BSP	.825	20.95	14	1.81	3/4	19
1/2 NPT	.8327	21.16	14	1.81		
3/4 BSP	1.041	26.44	14	1.81	61/64	24.5
3/4 NPT	1.05	25.53	14	1.81		
1" BSP	1.309	33.25	11	2.31	1 13/64	30.75
1" NPT	1.304	33.12	11½	2.21		

NPT (U.S. National Pipe Taper originally Briggs taper) has identical thread size and form to NPS (National Pipe Straight). Taper threads require PTFE tape to seal threads without galling; parallel or straight threads require washers to seal and should not be used with tape.

summary • **tape with taper threads**

Only 1/2 and 3/4 in NPT and BSP sizes are compatible.

LC in practice P A Bristow hetp, Macclesfield UK ISBN 0 950483 3 0 3

Compression fitting body threads

Compression fitting tube diameter in	Fitting body thread size in	threads per inch	thread type
1/16	10	32	UNF
1/8	5/16	20	NS
1/4	7/16	20	UNF
3/16	3/8	20	
3/8	9/16	20	
1/2	3/4	20	
3/4	1	20	
1	1 5/16	20	

10.7 Meshes and Sieving

Meshes

Single equal warp and weft meshes are used to sieve particles. Double weave meshes are suitable for retaining column packing materials and have a low impedance to liquid flow.

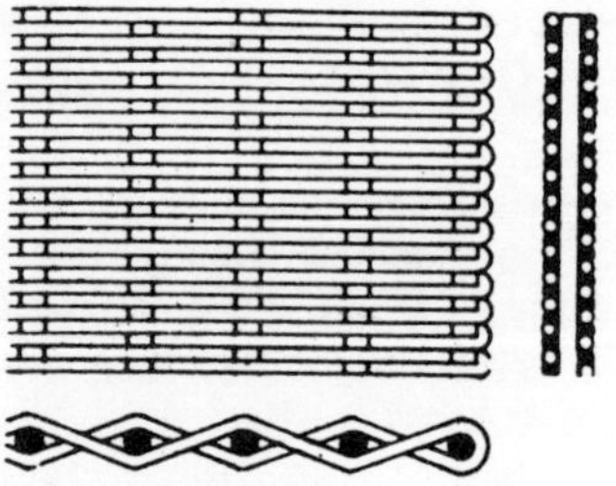

Dutch double-weave mesh

Source: Sankey-Greening Wire Weaving Co.
Thelwall Green, Warrington

Wire mesh—selected sizes

wires/in	wires/mm	wire diameters thou or mil	µm
325 x 2300	13 x 90	1.4 x 1	56 x 40
250 x 1400	10 x 56	2 x 1.4	50 x 35
165 x 800	6.5 x 31	3 x 1.4	120 x 56
165 x 1400	6.5 x 55	2.4 x 1.4	96 x 56
80 x 700	3.2 x 28	4 x 3	160 x 120
16 x 96	6.3 x 38	16.4 x 11.6	650 x 460

Sieves

Sieves are described in ISO/R 565, BS 410:1969, ASTM E11-61 and DIN 4188. Metric sizes dimensioned in µm aperture are now replacing the old sieve number based on the number of wires per inch. A table follows relating mesh number to aperture, that is the maximum size of particle that can pass.

Nominal aperture μm	Mesh number	Nominal aperture μm	Mesh number
1000	**16**	**180**	**85**
850	18	150	100
710	**22**	**125**	**120**
600	25	106	150
500	**30**	**90**	**170**
425	36	75	200
355	**44**	**63**	**240**
300	52	53	300
250	**60**	**45**	**350**
212	72	38	400

NOTE. The sizes shown in bold type are the principal sizes of ISO/R 565 with a ratio of about 1.4. The sizes taken continuously are in a series with a ratio of about 1.19.

Below 50 μm round-hole micro-plate sieves are made. These are perforated electrolytically. Available sizes are 50, 40, 30, 25, 20, 15, 10 and 5 μm.

Sieving

Sieving is not as useful as might be expected in sizing LC packings because the most interesting microparticles below 50 μm fall slowly and electrostatic charges attract them to each other and the sieve. In general it is impractical to separate more than a gram or two of material below 50 μm and the sizing process rapidly becomes less efficient as the size falls to 10 μm. Wet sieving is often more satisfactory but filtration and recovery of solids awkward. BS 1796:1952 describes methods and problems in using fine mesh sieves.

10.8 Sintered Materials

Stainless steel, glass and other ceramics, and PTFE can all be used to create sintered materials with a small pore size. BS 1752:1963 describes these materials and their characterisation by permeability and maximum pore size.

Sinters are usually thicker and therefore more rigid than woven meshes, but they also have a higher impedance to flow, mainly because they are thicker. Either factor may, or may not, be significant.

Pore diameter

The mean pore diameter is not readily determined but BS 1752 describes in detail a method for the maximum and uniformity of pore diameter. The pressure required to force a bubble of air through a wetted filter is measured to find the largest pore and the evenness of subsequent bubbles formed shows the pore uniformity. For pore diameters <25 μm, the maximum pore diameter is approximately $30\gamma/p$

where γ is the test liquid surface tension in mN/m and p the pressure in mmHg. Typical values are: water 72.8; methanol 22.6; carbon tetrachloride 27.0.

This test is probably pessimistic when predicting the smallest particle the sinter will retain. When this test is applied to a woven mesh material 325 x 2300 wires/inch the result is 8 μm; but in practice packings down to 1 μm are retained. (Strictly the Standard does not mention woven material but there seems no reason against its widespread use in this way.)

The following grades are described:

grade no	range of maximum pore diameter (μm)
0	150–250
1	90–150
2	40–90
3	15–40
4	5–15
5	⩽2

Permeability

For laminar flow through a sintered material (woven material or bed of small particles) the permeability is given by

$$K^\circ = \frac{z\eta q}{A\Delta p}$$

where K° = permeability (m^2)
z = thickness (m)
Δp = pressure differential (N/m^2)
η = viscosity (Ns/m^2)
A = area (m^2)
q = flowrate (m^3/s)

Thus it is quite easy to calculate the permeability from a few simple measurements.

Relationship between pore diameter and permeability

From Kozeny type equations the relationship

$$K^\circ \propto f d^2$$

where K° = permeability
f = fraction of void space
d = maximum pore diameter

is shown to hold for similar materials.

This can be used to predict the effect of changing porosity and pore diameter either in filters or sinters used to retain column packings, or in packed columns. Since it is difficult to make stable beds or sinters with very widely differing fraction void space, the particle or pore diameter dominates the observed permeability. Ideally the bed retaining element should not contribute to the overall column impedance. Assuming similar but smaller pore size for the sinter compared to a bed of particles, then the thickness should definitely be less than 1/100 of the column length. In practice woven mesh discs of 150 μm thickness and 8 μm nominal porosity have been found ideal.

10.9 Tools

A few tools are essential even with a manufacturer's complete instrument (and some will come equipped with such a tool kit).

Spanners are absolutely essential. It is possible to manage analytical columns with two 200 mm (8 in) vanadium steel adjustables but a wider size range is better. In theory fixed spanners are less likely to lead to spoilt fittings, but most users find adjustable spanners more convenient because so many fixed sizes are needed. The fact that adjustable spanners are usually wider reduces the risk of damage if the spanner is not adjusted to fit the nut tightly. Very large spanners are essential for 1 inch fittings and a long arm single-ended type gives good leverage. A useful selection would be: two 100 mm (4 in), two 200 mm (8 in), two 300 mm (12 in) and for preparative work two 500 mm (18 in) adjustable spanners.

Many instruments and components use flanged PTFE tube fittings. It is convenient to be able to remake these flanges (and connect different components). Flanging kits are available from most manufacturers of this type of fitting. Most are based on electrical heating of a metal former against which the tube is pressed; the Laboratory Data Control FT-1 is typical. Another version from **hetp** (part no 363) is based on a Curie point soldering iron bit which heats up very quickly to a controlled 380°C. A double-vee tube holder is needed to hold the tube and cold metal plate to cool the flange rapidly and set it in shape.

To cut and trim stainless steel tube to go into compression fittings a hacksaw, file and scraper are needed. An all-hard blade deals with this tough steel and a fine pitch (32 teeth per inch) cuts even thin wall tube without snatching. Small 'Junior' hacksaws are less satisfactory. A fine file will remove burrs on the outside of the tube but a triangular scraper is more effective on the inside. Needle files or a silicon carbide stone will be needed for 1/16 in tube. A needle will clear the centre hole which can become completely blocked after filing very small bore tubes. Column tubes with ends machined in a lathe or ground square are available from many manufacturers.

It is also possible to remove burrs chemically with aqua regia for a minute or so[142]. Electropolishing is an even more sophisticated method[112].

A small vice is always useful. Silicon carbide paper (400 and 600 mesh grades) and iron oxide (crocus) paper or diamond paste may be needed to repair damage to sliding seals in valves. PTFE tube can be cut with a scalpel or razor blade. Other special tools may be supplied for repairs or adjustments to pumps and detectors. Screwdrivers (slotted and Posidriv or Phillips heads) and hexagon head keys or drivers are commonly needed.

11 DATA TABLES

11.1 Notes on Format

SI units are used wherever reasonable adopting recommendations of ISO/R31 and BS 3763 on symbols and conversion factors. Many physical constants have been converted to SI form using BS 350: Pt 1: 1974. This sometimes leads to some unfamiliar units but should make calculations much easier and thus encourage use of the data. The use of typical data in SI form and a collection of conversions factors drawn from BS 350 should also help. Notification of any conversion errors will be welcomed.

Nominal fractional inch sizes will persist for a long time especially in pipe sizes: these have been retained to avoid ugly and unhelpful conversions to metric units. Thus tubes are described as 1/4 in outside diameter but 5 mm bore.

Chromatographers have failed to 'go metric' in persisting with psi and minutes, both of which impede quick mental calculations by introducing unnecessary conversion factors and are no more convenient than the corresponding MPa or bar, and μl/s or ml/s.

11.2 Abbreviations

LC	liquid chromatography
GC	gas chromatography
od	outside diameter
id	inside diameter (bore)
LLC	liquid–liquid chromatography
LSC	liquid–solid chromatography
HPLC	high pressure LC
HSLC	high speed LC
HELC	high efficiency LC
NMR	nuclear magnetic
MS	mass spectrometry
UV	ultraviolet (spectrometry)
UV-VIS	ultraviolet-visible
RI	refractive index
RF	radio frequency
PTFE	polytetrafluoroethylene
PCTFE	polychlorotrifluoroethylene
ϕ	diameter symbol
PLB	porous layer bead
IEC	ion-exchange chromatography
LSAC	liquid–solid adsorption chromatography
PC	permeation chromatography
GPC	gel permeation chromatography
prep	preparative
ODS	octadecyl silane
HETP	height equivalent to a theoretical plate
TLC	thin-layer chromatography

11.3 Conversion Factors

Units of quantities commonly used in LC are defined in terms of their base SI unit (combination of m, kg, s, mol, K). To convert the non-SI unit to base unit value for calculation, multiply by the factor given. Data taken mainly from BS 350:part 1: 1974. SI abbreviations conform to recommendations of BS 3763:1970.

Length metre (m)
1A = 0.1 nm = 10^{-10} m
1 thou or mil = 10^{-3} inch = 25.4 μm = 0.025 mm = 25.4 x 10^{-6} m
1 microinch (μ in) = 10^{-6} inch = 25.4 nm = 0.025 nm = 25.4 x 10^{-9} m

Area (m^2)
1 mm^2 = 10^{-6} m^2
1 μm^2 = 10^{-12} m^2
1 nm^2 = 10^{-18} m^2 = 100Å^2

Volume (m^3)
1 μl = 1 mm^3 = 10^{-9} m^3
1 ml = 1000 μl = 1000 mm^3 = 10^{-6} m^3
1 litre = 1000 ml = 10^6 mm^3 = 10^{-3} m^3

Time second (s)
1 min = 60 sec
1 hr = 3600 sec

Linear Velocity (m/s)
1 mm/s = 10^{-3} m/sec
1 mm/min = $10^{-4}/6$ = 1.67 x 10^{-5} m/s
1 μm/s = 10^{-6} m/s

Volumetric Flowrate (m^3/s)
1 l/s = 0.001 m^3/s
1 μl/s = 10^{-9} m^3/s
1 l/min = 16.7 ml/s = 1.67 x 10^{-5} m^3/s
1 ml/min = 16.7 μl/s = 0.0167 ml/s = 1.67 x 10^{-8} m^3/s

Specific Surface area per unit mass (m^2/kg)
1 m^2/g = 1000 m^2/kg
1 mm^2/mg = 1 m^2/kg

Concentration (kg/m^3)
1 kg/m^3 = 1 g/l = 1 mg/ml = 1 μg/μl

Force newton (N or kg m/s^2)
1 dyne = 10^{-5} N
1 kp = 1 kgf = 9.80665 N

Pressure pascal (N/m^2 or Pa)
1 $dyne/cm^2$ = 0.1 N/m^2 = 0.1 Pa
1 kgf/cm^2 = 0.980665×10^5 N/m^2 = 0.0980665 MPa = 1 at = 1 kp/cm^2
1 lb/in^2 (psi) = 6894.76 Pa or N/m^2
1 in H_2O = 249.089 Pa $\simeq$0.0025 bar; 1 m H_2O $\simeq$0.1 bar
1 mm Hg = 1 torr = 133.322 Pa; 1 mm H_2O $\simeq$10 Pa
1 bar = 10^5 N/m^2 or Pa = 0.1 MPa or MN/m^2 or N/mm^2

Stress (N/m^2 or Pa)
1 N/mm^2 = 1 MN/m^2 = 1 MPa
1 kgf/mm^2 = 9.80665 N/mm^2
1 kgf/cm^2 = 0.0980665 MN/m^2

Density (kg/m^3)
1 g/ml = 1000 kg/m^3

Viscosity (dynamic) (Pa s or Ns/m^2 or kg/(ms))
1 Poise = 10^{-1} Ns/m^2 = 10^{-1} Pa s
1 cP = 10^{-2} = 10^{-3} Pa s = mN s/m^2 = mPa s

Viscosity (kinematic) (m^2/s)
1 St = 1 cm^2/s = 10^{-4} m^2/s
1 cSt = 10^{-2} = 10^{-6} m^2/s (=1 mm^2/s)

Energy (J)
1 J = 1 Nm = 1 Pa m^3
1 erg = 10^{-7} J
1 kWh = 3.6 MJ
1 Litre-atmosphere = 101.325 J

Power (W or J/s)
1 hp = 745.700 W
1 cal/s = 4.1868 W

Specific Heat Capacity (or Specific Entropy) (J/(kg K))
1 kcal/(kg K) = 4186.8 J/(kg K)

Thermal Conductivity (W/(m K))
1 W/(m°C) = 1 W/(m K)
1 kcal/(mhr K) = 1.163 W/(m K)
1 cal/(Cm s K) = 418.68 W/(m K)

Diffusivity (m^2/s)
1 cm^2/s = 10^{-4} m^2/s

Specific Permeability (m^2)
1 cm^2 = 10^{-4} m^2 = 10^8 μm^2 1 darcy = 9.869×10^{-9} cm^2 = 9.869×10^{-13} m^2

Compressibility (m^2/N)
1 at^{-1} = 1.020×10^{-5} m^2/N

11.4 Photometry Units and Conversion Factors

Quantity		SI units
transmittance $T = I/I_o = 10^{-A}$ = antilog (-A) = $1/10^A$ (Defined in fractional terms: percentage also often used)		dimensionless
absorbance $A = \log_{10}(1/T) = \epsilon cl$		dimensionless
specific absorbance[240] (extinction coefficient)	$\epsilon = A/lc$	m^2/kg
specific molar absorbance $\epsilon^{mol/l}_{1\ cm} = 10^5 \times \epsilon\ m^2/mol$		m^2/mol
concentration c	1 mol/l = 10^3 mol/m^3 1 g/l = 1 kg/m^3	kg/m^3 or mol/m^3
pathlength l		m

Absorbance to Transmittance Conversion

A	T
0	1.000
0.00005	0.999885
0.0001	0.99977
0.001	0.9977
0.005	0.9885
0.01	0.977
0.05	0.891
0.1	0.794
0.5	0.316
1	0.1
2	0.01
3	0.001
4	0.0001

T	A
1.0	0
0.9999	4.34×10^{-5}
0.9995	2.17×10^{-4}
0.999	4.34×10^{-4}
0.995	2.17×10^{-3}
0.99	4.36×10^{-3}
0.95	2.23×10^{-2}
0.9	4.58×10^{-2}
0.8	9.69×10^{-2}
0.7	0.15
0.6	0.222
0.5	0.301
0.4	0.398
0.3	0.523
0.2	0.698
0.1	1.0
0.01	2.0

11.6 Symbols

Symbols are chosen to conform (as far as compatible) with BS 1991: Pt 2: 1967; IUPAC Recommendations on Nomenclature for Chromatography[177]; and Giddings[1].

A	eddy diffusion term in plate height equation
C	mass transfer term in plate height equation
D_m	molecular diffusion coefficient (diffusivity) of solute in mobile phase
d	diameter
d_c	diameter of column
d_p	diameter of particle
E	separation impedance
ϵ	total porosity
ϵ_I	interparticle porosity (interstitial) = V_I/X
F	flowrate (linear in tubes)
F_c	volumetric flowrate
H	plate height or height equivalent to a theoretical plate (HETP)
h	reduced plate height
I	Knox-Parcher ratio
K	distribution or partition coefficient (stationary phase concentration/mobile phase concentration)
k'	column capacity factor = $(t_r - t_o)/t_o$, *mass* distribution ratio
K^o	specific permeability
K_o	distribution coefficient for permeation chromatography
L	length of column
l	reduced column length
N	plate number
P	pressure (usually an implied pressure drop or difference)
$r_{A/B}$	relative retention = V_R^A/V_R^B
R	retention ratio = $1/(1 + k')$
R_s	resolution
R_f	retention ratio for TLC
Re	Reynolds number
t	time
t_r	retention time (component may be indicated by superscript)
t_o	time for non-sorbed peak
u	interstitial or linear velocity
V_R	retention or elution volume of solute (perhaps identified by superscript)
$V_R{}^o$	specific retention volume
V_M	total volume of mobile phase in column including pores = $t_o F_c$
V_I	interstitial volume excluding pores
V_P	volume of pores = $V_M - V_I$
w	width of zone (4σ) in various units
w_i	width of injected zone
γ	obstruction factor for diffusion through granular materials
ϵ^o	solvent strength parameter
η	viscosity
ν	reduced velocity
ρ	density
σ	standard deviation of zone (subscripts show units of length, time, volume)
ϕ'	flow resistance parameter

11.7 Glossary of Terms used in Chromatography

(Drawn mainly from the IUPAC recommendations[177])

Adsorption chromatography	–	separation based mainly on differences between adsorption affinities for the surface of an active solid.
Affinity chromatography	–	very specific form of bonded phase which adsorbs only one specific class of molecules.
Band	–	synonymous with zone, where the solute is situated.
Band spreading	–	increase in the distance, time or volume containing the solute.
Bonded-phase	–	layer of organic material chemically bonded to the surface of the pores.
Chromatography	–	separation system where a fluid moves with respect to a solid and the sample components distribute between the two phases.
Column capacity factor (or ratio) k'	–	measure of degree of retention of a solute, compared to a non-sorbed compound.
Displacement chromatography	–	elution procedure where the eluent contains a compound more retained than the components of the sample.
Distribution coefficient (partition coefficient)	–	ratio of concentration in stationary phase to that in mobile phase.
Eluate	–	effluent from a chromatographic bed when elution is carried out.
Eluent	–	solvent which elutes components from bed.
Eluotropic series	–	group of solvents placed in order of their eluting strength.
Elution chromatography	–	eluent is passed through chromatographic bed after sample is applied.
Elutriation	–	process used to size particles by sedimentation.
Exclusion	–	see permeation chromatography.
Exclusion limit	–	molecular size which is just excluded.
Flowrate (volumetric)	–	volume per time passing through the column.
Flow programming	–	changing the flowrate during the elution.
Frontal chromatography	–	sample is fed continuously into the chromatographic bed.
Fronting	–	peak asymmetry, the opposite of tailing.
Gel filtration	–	see permeation chromatography.
Gradient elution	–	continuously changing the composition of the eluent during elution.
Ion-exchange chromatography	–	separation based mainly on affinities for ion-exchanging groups.
Linear capacity	–	mass (over)load which does not change retention parameter or k' by more than certain amount, usually 10%.
Linear (interstitial) velocity	–	average speed at which an eluent molecule travels down the column.
Macro-reticular	–	particle in which some pores are much larger than those in which most separation takes place.

Mass-transfer	–	process of equilibration of solute molecules between mobile and stationary zones, essentially liquid outside and inside pores.
Mobile phase	–	fluid which moves with respect to the solid chromatographic bed, usually called eluent. Liquid outside pores is often implied.
Molecular exclusion	–	see permeation chromatography.
Partition chromatography	–	separation based mainly on differences in solubility.
Peak capacity	–	number of peaks that could, if equally spaced, be separated on a column for a given column capacity factor k'.
Pellicular	–	column packing material with a solid core and a thin porous skin.
Permeation chromatography	–	separation based mainly on differences in size or shape which lead to varying degrees of exclusion.
Plate height	–	measure of the amount of zone broadening a column causes compared to its length. (Also called height equivalent to a theoretical plate or HETP.)
Plate number	–	measure of potential separating power of a column. Determines the peak capacity.
Reduced parameter	–	dimensionless number which is corrected for particle size.
Resolution	–	degree of separation between two solutes. The difference between retention parameter (volume, time, distance) compared to the mean retention parameter.
Reversed-phase chromatography	–	system where stationary phase is non-polar, and eluent is polar.
Salting-out chromatography	–	adding a non-sorbable electrolyte to the eluent to increase the polarity of the eluent, thereby reducing the solubility of organic molecules in the eluent.
Separation factor	–	ratio of retention parameters for two solutes (time, volume, or distance). (By convention >1)
Sizing	–	see permeation chromatography.
Solvent	–	a liquid, often used as an eluent.
Solvent demixing	–	non-equilibrium state where one component of the eluent is selectively being absorbed.
Standard deviation	–	about 1/4 of peak base width. Can be in units of time, distance, volume or chart distance.
Stationary phase	–	material on which components must be retained to achieve separation; column packing material. (Often not strictly a phase, but a zone or layer.)
Stepwise elution	–	changing the composition of the eluent during elution.
Superficially porous	–	see pellicular.
Thin-layer chromatography (TLC)	–	chromatography carried out on a thin layer of adsorbent spread on a support like a glass plate.
Variance	–	measure of zone broadening. (Equal to square of standard deviation, or base width^2/16.) Variances due to injection, column and detection system are additive.
Zone	–	zone of volume, time, or distance where the solute is distributed.

11.8 Calculation of Plate Number

Table for Calculation of Column Efficiency

Ratio	Plate No.	Ratio	Plate No.
1	5	56	17370
2	22	57	18000
3	50	58	18640
4	90	59	19280
5	140	60	19940
6	200	61	20600
7	270	62	21000
8	350	63	22000
9	450	64	22700
10	554	65	23400
11	670	66	24100
12	800	67	24900
13	936	68	25600
14	1090	69	26400
15	1250	70	27200
16	1420	71	27900
17	1600	72	28700
18	1790	73	29500
19	2000	74	30300
20	2210	75	31200
21	2440	76	32000
22	2680	77	32900
23	2930	78	33700
24	3190	79	34600
25	3460	80	35500
26	3750	81	36400
27	4040	82	37300
28	4340	83	38200
29	4670	84	39100
30	4990	85	40000
31	5320	86	41000
32	5670	87	41900
33	6030	88	42900
34	6400	89	43900
35	6790	90	44900
36	7180	91	45900
37	7580	92	46900
38	8000	93	47900
39	8430	94	49000
40	8860	95	50000
41	9310	96	51100
42	9770	97	52100
43	10240	98	53200
44	10730	99	54300
45	11220	100	55400
46	11720	105	61100
47	12240	110	67000
48	12860	115	73300
49	13300	120	79800
50	13849	125	86600
51	14409	130	93600
52	14980	135	101000
53	15560	140	109000
54	16150	145	117000
55	16750	150	124000

$$\text{ratio} = \frac{\text{retention time}}{\text{width at half ht}}$$

12 MISCELLANY

12.1 Fault finding

summary • **diagnosing faults is easier in modular instruments**

Diagnosing faults can be difficult because faults in one part, for example the pump, may only manifest themselves in another part like the detector. It is obviously important to pin-point the real source of the problem.

It is much easier to find faults with modular instruments where the pumping system, column, and detector can be interchanged, particularly if two detectors and recorders can be coupled up simultaneously as a temporary measure.

Summary of likely sources of observed symptoms

Symptom	Likely source of problem (most probable first)
Flowrate incorrect	pump; blockage in pipework or column; leaks.
Change of peak retention time	column; eluent; flowrate (pressure).
Peak shape	injection problem; too large volume or mass injected; chromatographic difficulty; detector outside linear range.
Baseline	bubbles; solvent purity.
Noise	detector; pump pulsations; recorder.
Drift	detector; column; eluent composition changes.
No peaks	injection fault; chromatographic conditions wrong; detector; recorder.

Pumping system

Symptom	Suggested action
No flow or pressure	Check for obvious faults like motor not moving, air supply off. Check for airlocks in solvent feed, or air in system: disconnect column and pump, or recycle, at maximum rate until air is pumped through. Check pump valves for correct operation. A bubble of air introduced into the feed line is a useful diagnostic tool. If the bubble does not move then the inlet valve doesn't open, or the output leaks. If the bubble moves back and forth, the inlet valve leaks. It is easiest to swap the whole valve assembly and repair the faulty one at leisure.

If pressure builds up correctly, but there is no flow, check for a blockage by carefully loosening fittings working back from the detector until a leak is produced.

Too high pressure with constant flow pump, or too low flow with constant pressure pump — Locate partial blockages. Calculate expected pressure with length and packing material (see section 11.5)

Erratic pressure or flow — Check driving air or nitrogen pressure. Dirty valves can cause flow variations with reciprocating pumps. Airlocks, or dissolved air can also cause erratic flow.

Note that if the pressure is constant, then only a change in column, or pipework impedance can change the flowrate. But if the pressure and flow correlate, then the pump is the cause of the problem.

Column and injection system

If all peaks have the same form, for example showing tailing, then the injection system or column packing are nearly always the cause. Chromatographic factors can cause some peaks to be broader than others, or to tail. It is difficult to suggest general cures for necessarily specific chromatographic problems. The information below suggests how to improve the mechanical aspects of the column and injector that cause poor peak shape—which results in lower resolution.

Symptom — **Suggested causes and actions**

Broad peaks

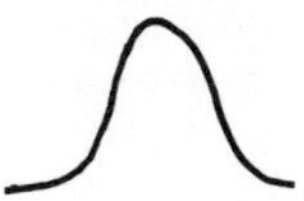

Poor packing (see section 4 for suggested packing methods). Non-infinite diameter column systems: to improve use a wide enough column with a central on-column injection.

Too large injection volume (unlikely unless deliberate).

Too fast—or too slow—flowrate (see section 3 for suggested flowrates for various sized columns).

Flat-topped peaks

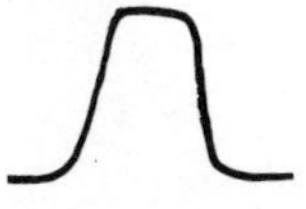

Detector being used outside linear response range (often seen under preparative conditions): remedy may be to change flowcell pathlength, or monitoring wavelength.

Too large injection volume (only seen when the injected volume is comparable with the column volume, say half the empty column tube volume).

Tailing

Usually caused by a *void* space in the column, or more commonly, the connection from the injection system to the top of the column, especially if this space is not well swept. Poorly designed injection valves also produce tailing.

Can be caused by *settlement* of the column packing (sometimes because the silica has dissolved in the eluent).

Too large a *mass* of sample can cause peak tailing, and a marked reduction in the retention time of the *front* of the peak.

Very rarely caused by the detector flowcell unless peak volumes are very small.

Fronting

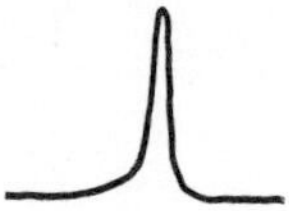

Usually attributed to sample reaching the *column wall*, where it travels more rapidly down the column than the bulk which goes down the central core of the column. Repacking the interface between the injector and the column (for example frits, glass beads or wool, or meshes) can sometimes cure the problem, but repacking the whole column may be necessary.

Detectors

If the detector is suspect, it is useful first to turn off the flow (and make sure by inspection that it really has stopped) and see if the symptoms persist. If the symptoms disappear then it is more likely that the source of the problem lies elsewhere. Duplicating or substituting detector and/or recorder is often the quickest way of producing an accurate diagnosis. For this purpose quite long lengths of 1/16 diameter ptfe tubing are acceptable. A dummy cell which has a suitable aperture, but without windows or liquid connections is a useful accessory to check that the detector is operating correctly and the lamp and optics are correctly aligned. An empty cell is a more expensive substitute.

Symptom — **Suggested action**

Won't zero

Check that the detector lamp is lit, and that the recorder zero is not offset too much.

Check for bubbles in sample or reference beams, or dirty windows. Collect a sample of the liquid coming from the detector and run a spectrum to check it has sufficient transparency for the wavelength in use. Ensure that the liquid in the reference beam is not contaminated, or decomposed by light, or absorbing too little compared to the sample beam. Leaks from the sample section can cause contamination of the reference section.

Some detectors will not zero if the lamp output is too low.

Noise

Bubbles, or droplets of an immiscible liquid are a very common cause of noise. Slight leaks cause air to leak *in*—not out, as might be expected. If restriction of flow downstream of the detector changes the detector reading immediately then there is a bubble.

Earth loops in connecting recorder to detector cause noise that persists regardless of attenuation settings. Try altering the signal grounding connections, but do not remove protective earth lines.

Spikes

If spikes persist when no liquid is in the cell, may be mains supply or RF borne. If mechanical shocks cause spikes, a loose connection should be suspected. Passing bubbles can cause spikes.

Try substituting lamps: deuterium lamps often become noisy before they fail completely.

Drift — Check first for solvent based problems: changing eluent composition, column bleed, very broad peaks, contaminants from pumps, pipework, septa etc. Check for slowly forming or disappearing bubbles, and leaks from sample to reference sections of the flowcell.

Cell windows can become dirty from leaking eluent, or from the air.

If drift persists with no liquid in the flowcell, then the detector lamp, or electronics is faulty.

Temperature changes can cause drift, especially with RI detectors.

Negative peaks — Check that the sample reference electronics and pipework are correctly connected. Check that the leads to the recorder are connected with the right polarity.

Derivative peaks — These are caused by large changes in the refractive index of liquids passing through photometric detectors.

12.2 Peak Shape in LC

summary

- **peak shape important for good resolution**
- **injection mainly determines peak shape**
- **importance of wall effects on peak shape not clear**
- **larger diameter columns(5-10 mm bore) often give best results**

In GC, peaks of truly symmetrical Gaussian form are rare: in LC they are even rarer because there are more possible causes why peaks should not have the form predicted by theory. This is important in practice because a small amount of tailing or fronting has a marked effect on the resolution that can be attained. Some of the factors affecting peak shape in GC have been studied by Ettre and co-workers[201] concentrating on the injection system, the main culprit. In GC the diffusivity of the gas is so high that molecules interchange position from wall to column centre many times as the solute zones pass down the column. In typical LC columns, where the diffusivity of the eluent is a thousand times slower, few molecules diffuse to and from the walls during the time the chromatogram is run. Thus an averaging effect in GC is absent in many LC column configurations.

Thus the distribution of the injected sample and its interaction with any wall effects distort the peak in a complex way. Asymmetric broadening caused by the column end fittings and detector flowcell provide a further complication. Overall, however, the profile of the injected slug is reflected quite precisely by the detector if the zone broadening of the column and a slight tailing caused by the detector is taken into account.

On top of all these factors which cause asymmetric peaks using an 'inert particle' column, there are some rather more chromatographic reasons for asymmetry. A non-homogeneous stationary zone is the most common cause. A chromatographic system (column, eluent and solute) behaves as if packed with a mixture of materials which adsorb differently. The surface itself can be non-homogeneous with groups of different retaining 'activity' – though in most systems undoubted variations in the surface are hidden by layers of adsorbed molecules, usually the polar component of an eluent mixture. Solutes can appear to be non-homogeneous because some injected molecules are ionised, or conversely only some may be converted to ion-pairs by the eluent: in this way one injected compound can give two peaks, a most puzzling phenomenon. Mass overload can also cause a distortion of peak shapes, and reduction in retention.

It is usually possible to distinguish these two possibilities by noting that flow geometry causes all peaks to be affected in a similar way (though tempered by concentration-by-adsorption on the top of the column, and by steadily increasing peak volumes for later peaks), whereas different compounds usually react differently to inhomogeneity in the stationary or mobile zones, so some peaks are more tailed or broadened than others.

For dry-packed columns at least, wall effects have been shown to play an important part in distorting the peak shape[120,221]. Knox and co-workers have used the concept of the 'infinite diameter' column[106] to make wall effects negligible by using such a wide column that solute molecules injected at the centre never reach the walls by slow radial diffusion.

Columns of microparticles packed from slurries have not yet been studied by the rigorous methods used by Knox[120] to study dry-packed columns. The many results by many different groups, obtained by observing overall peak efficiency and peak shape alone, can only be described as contradictory. Further work will be required to resolve this controversy about the importance of wall effects.

In practice two conclusions are reasonably clear:

- Injection plays a critically important part in determining peak shape, and thus the observed efficiency, and resolution (much more than in GC[201]).
- Larger diameter columns (at least 4 mm bore, preferably >5-10 mm) give better results in many workers' laboratories. This may be a result of the larger volumes that can be injected, the larger peak volumes that result, reduced wall effects from the use of columns in the infinite diameter mode, or simply that the proportion of material that is near to the wall (say within 30 particle diameters[120]) is smaller for a larger column diameter.

12.3 Radial Dispersion

summary
- **radial dispersion important because of much lower efficiency near column walls**
- **column operates in 'infinite diameter' mode if solute does not reach wall**
- **for microparticles (<20 μm) zone spreads several mm at most efficient speed**
- **injection distribution determines width in practice**
- **large pellicular packings (~40 μm) in long columns average out wall effects and are insensitive to injection distribution**
- **for 5 μm particles, 100 mm length must be >5 mm bore to avoid wall effects with certainty**

Knox and his co-workers[120] have studied the radial diffusion of solute in LC. They found that near the wall much more zone broadening takes place than in the main body of the column. It may be useful to avoid this wall zone and thus operate in the 'infinite diameter' mode. From their experimental results on dry-packed spherical glass beads ~60 μm diameter they derive the following equation for h_r, the reduced radial plate height

$$h_r = 1.4/\nu + 0.06$$

The radial plate height $H_r = h_r r d_p$ can be calculated from

$$H_r = 1.4\, d_p/\nu + 0.06\, d_p$$

Substituting values as follows: $D_m = 3 \times 10^{-9}\ m^2/s$; $d_p = 5 \times 10^{-6}$; $l = 0.1$ m; and t_o, the eluent time, as 30 s, gives a value of

$$\nu = L\, d_p/D_m\, t_o = 5.5 \qquad \text{and} \qquad H_r = 1.3 \times 10^{-6} + 0.3 \times 10^{-6} = 1.6\ \mu m.$$

The radial variance $\sigma_r^2 = H_r L$ and the standard deviation $\sigma_r = \sqrt{H_r L}$, and thus the diameter of the zone at the column end $w_r = 4\sqrt{H_r L}$, and the distance a negligible volume injection spreads sideways is $2\sigma_r = 2\sqrt{H_r L}$.

With the above values this distance is $2 \times \sqrt{1.6 \times 0.1} = 0.8$ mm, and the diameter at the column end 1.6 mm.

This diameter can also be calculated directly from the experimental measurements of D_m, d_p, L, and t_o by rearranging the equation to give

$$w_r \simeq \sqrt{20\, D_m\, t_o + L\, d_p}$$

If the following units are chosen: D_m (mm/s), L (m), d_p (μm), t_o (s), then w_r is in mm. Thus with the above values

$$w_r = \sqrt{20 \times 3 \times 10^{-3} \times 30 + 0.1 \times 5} = \sqrt{1.8 + 0.5} = \sqrt{2.3} = 1.5\ \text{mm}$$

If we take the injection width w_i into account then the

$$\text{total radial width } w_t = \sqrt{w_r^2 + w_i^2}$$

If we assume that the liquid from a syringe needle spreads to a sphere of diameter d from a volume V, then $d^2 \simeq 1.6\ V^{0.67}$. (This is unlikely to describe an injection distribution very precisely, but the calculated width is only slightly affected by assumptions about its geometry provided it is a central on-column injection. It also assumes that the pores are available to the injection slug which may not be true for high speed injections).

Width of injected sample – assuming spherical distribution

V (μl)	$V^{0.67}$	w_i^2	w_i
1	1	1.6	1.3
2	1.6	2.5	1.6
5	2.9	4.7	2.2
10	4.7	7.4	2.7
20	7.4	12	3.5
50	13	22	4.7
100	22	35	5.9

For a 10 μl injection $\quad w_i^2 = 1.6 \times 10^{0.67} = 1.6 \times 4.6 = 7.4\ \text{mm}^2$

and from the above calculation $w_r^2 = 2.3$, therefore

$$w_t^2 = \sqrt{2.3 + 7.4} = \sqrt{9.7} = 3.1\ \text{mm.}$$

Allowing for the injection not being exactly in the centre of the column, we can still expect a 5 mm bore column to operate in the 'infinite diameter' mode. (But note that the injection distribution is the major contributor to the radial width.)

The overall practical equation for w_t is thus:

$$w_t = \sqrt{20\ D_m\ (\text{mm}^2/\text{s})\ t_o(\text{s}) + L\ (\text{m})\ d_p\ (\mu\text{m}) + 1.6\ V\ (\mu\text{l})^{0.67}}$$

If we calculate the situation for a long column packed with a pellicular material, taking

$$D_m = 10^{-3}\ \text{mm}^2/\text{s};\ t_o = 100\ \text{s};\ L = 1\ \text{m};\ d_p = 40\ \mu\text{m};\ V = 10\ \mu\text{l then}$$

$$w_t = \sqrt{20 \times 10^{-3} \times 100 + 40 + 7.4} = \sqrt{49} = 7\ \text{mm.}$$

Thus if the bore is only a few mm, then the injection will spread to and from the walls and any wall effects will be well averaged. The column should give good peak shapes but with lower efficiency in terms of reduced plate height than an infinite diameter column. The injection volume and distribution then make only a negligible contribution to the overall radial width.

Columns of intermediate geometry tend to give peak shapes which vary as slightly more or less material reaches the wall effect zone. This can lead to problems with quantitation.

12.4 Connecting Tubes, Fittings, and Flowcells —zone broadening, pressure drop and velocity

Zone Broadening

summary
- **for analytical columns use ~ 0.3 mm bore (12 thou) <200 mm long**
- **above 200 mm length use 0.2 mm bore (up to ~1 m)**
- **for prep use ~0.8 mm bore (30 thou) <1 m**
- **σ_{tube} <1/10 peak volume**

Separated peaks should be re-mixed as little as possible in the connecting tubes, or in the detector and connecting fittings. Samples once injected should not be dispersed and diluted in the pipework or injection valve. This section gives examples of calculations and some derived practical conclusions. Most calculations tend to be approximate, especially when complex flow paths like those in detector flowcells and injection valves are involved, and experimental verification is desirable[103].

Ideally broadening caused outside the column should be negligible compared to the broadening function of the column packing itself. As materials have become smaller and packing more efficient this has become more difficult to achieve in practice.

Calculation is based on the additivity of variances (σ^2). (See Section 3).

Thus $\sigma^2_{total} = \sigma^2_{column} + \sigma^2_{tube}$

If a peak (and usually it is sensible to choose the smallest, probably unretained, peak) is to be broadened by 10% by the connecting tube

Then $\sigma_{total} \leqslant 1.10\ \sigma_{column}$ and $\sigma^2_{total} \leqslant 1.21\ \sigma^2_{column}$

$\sigma^2_{tube} \leqslant 0.21\ \sigma^2_{column}$ and $\sigma_{tube} \leqslant 0.46\ \sigma_{column}$

The 'standard deviation' σ can be assessed by calculation, or from the observed peak width (4 σ_{column}) in μl from a chromatogram.

A good target for tube broadening is 0.46 x peak width/4 or <1/10 peak width.

For example, a short efficient column (100 mm long, 5 mm bore) and plate number 3600 might have an unretained peak volume of 100 μl.

Thus $\sigma_{column} \approx 25\ mm^3$ and $\sigma^2_{column} \approx 625\ mm^6$,

so we should aim for $\sigma_{tube} < 12\ mm^3$ and $\sigma^2_{tube} < 144\ mm^6$,

or from the rule of thumb $\sigma_{tube} < 1/10 \times 100\ \mu l$ or 10 μl.

Dispersion caused by tubing can be calculated from the Taylor-Golay equation (Note that experimental evidence[227] suggests that this equation should be applied with caution when predicting peak broadening in LC pipework. In practice asymetric peaks are common.)

The equation can be expressed as

$$\sigma^2_{tube} = \frac{d^4 F z}{122 \times D_m}$$

where
d = bore (mm)
F = flowrate (μl/s)
z = length of tube (mm)
D_m = diffusivity (mm^2/s)
σ^2 = peak variance (mm^6)

Taking F as 60 μl/s and D_m as 3.5×10^{-3} mm^2/s typical of solvents like hexane, and requiring $\sigma^2_{tube} \leqslant 144$ mm^6, then approximately

$$z_{max} \leqslant 1/d^4$$

Note that the permissible length depends critically on the internal diameter. For this column under these conditions the maximum length is calculated below[103].

bore mm	thou	max. length mm	σ per m mm^3 or μl	σ^2 mm^6
0.1	4	2500	3.7	14
0.15	6	500	8.4	71
0.2	8	150	15	224
0.3	12	30	34	1140
0.5	20	4	60	3600

Obviously nominal 12 thou bore is undesirable except in very short lengths, but half a metre of 6 thou tube is quite safe. (A quick go—no go test for tube is to test whether a microlitre syringe needle (nominal diameter 0.5 mm) will fit into the bore: if it will the tubing is definitely unsuitable for analytical column to detector connections.) Larger, less efficient, or longer columns, or different conditions, may allow other diameters and lengths to be used without resolution loss. The following table gives selected values for flowrate of 60 mm^3/s and length 1 m, taking D_m as 3.5×10^{-3} mm^2/s.

Zone broadening in tubes

bore d mm	thou	d^4 mm^4	σ mm^3 or μl	σ^2 mm^6
0.1	4	10^{-4}	0.22	0.05
0.15	6	5×10^{-4}	0.3	0.1
0.2	8	1.6×10^{-3}	0.9	0.8
0.3	12	8×10^{-3}	2	4
0.5	20	6×10^{-2}	5.5	30
0.8	31	0.4	14	196
1.6	63	6.5	30	920

Broadening in Connections

Because of changes in tube dimensions which will lead to mixing and turbulence, the Taylor-Golay equation is hardly applicable to short larger diameter sections. However in an unbored reducing union 1/4—1/16 in with bore 0.052 in (1.32 mm) and length 3 mm, and under conditions of flowrate and diffusivity used in calculations above, one can calculate a very approximate value for the variance:

$$\sigma^2 = 1.32^4 \times 60 \times 3/(122 \times 3.5 \times 10^{-3}) = 1300 \text{ mm}^6$$

The predicted zone broadening is qualitatively confirmed in practice. Bored out fittings that do not have this void should obviously be used.

Broadening in Detector Flowcells

As in fittings, larger diameter flowcells are swept very inefficiently, but mixing may help to improve flushing. Thus for laminar flow the Taylor-Golay equation predicts $\sigma^2_{cell} = 1400$ mm^6 for a 10 mm long, 1 mm bore flowcell, which is large compared to σ^2_{column} of 700 mm^6.

If a flow F enters a fully mixed volume Q then the concentration of a peak will decline with time t by

$$c/c_o = \exp -(Ft/Q)$$

The variance introduced by this flushing can be shown to be

$$\sigma^2 = 2\,Q^2 \quad \text{or} \quad \sigma = \sqrt{2}\,Q$$

Since σ can be $\leqslant$12 mm^3, typical cell volumes of 8 mm^3 can just be tolerated if mixed.

In practice the resultant variance σ^2 will be between 1400 and 144 mm^6, values calculated for these two extreme hypotheses. Nevertheless improvements in cells are clearly needed even for 5 μm particles packed efficiently into small diameter columns.

Values calculated from the Taylor-Golay equation for other possible flowcell configurations are:

diameter mm	length mm	aperture mm^2	volume mm^3 or μl	σ mm^3	σ^2 mm^6
0.8	5	0.5	2.5	17	293
0.5	5	0.2	1	6.5	42
1	10	0.8	8	37	1400
1	1	0.8	0.8	12	140
1.5	10	1.8	18	84	7000
2	1	3	3	47	2200
3	10	7	70	330	110000

Longer and thinner cells give least zone broadening and highest sensitivity for a given volume, but reduce the available light.

Experimental data is sparse[107,108,239] and determination should now form part of every detector specification.

Pressure Drop in Tubes

summary
- **~0.2 mm bore tube gives negligible pressure drop for analytical columns**
- **~0.4 mm bore better for columns above 5 mm diameter**
- **~0.8 mm bore, 1/16 diameter gives pressure drop of less than a few bars even for 1 inch prep columns**

Reynolds number Re is given by linear velocity x diameter/viscosity. Even at very high flow rates, suitable for 100 mm diameter columns, say 10 ml/s flowing in 1/8 in diameter, 1.6 mm bore tube, Re is still only about 100 and for analytical conditions is 1. Only when Re is 10^3 does flow become turbulent, so flow in tubes may always be assumed to be laminar in LC.

From Poiseuille's formula, valid for laminar flow, $\Delta p = (8/\pi)\,F\,\eta\,z/r^4$

some pressure drops are calculated where the viscosity η is taken as 10^{-3} Ns/m^2 (but could be half this value for many organic solvents), and the length z as 1 metre.

tube diameter inch	bore mm	flowrate μl/s	pressure drop bar
1/16	0.1	10	40
		1000	4000
	0.2	10	2
		1000	200
	0.4	100	2
	0.5	10	0.1
		1000	10
	0.8	1000	1
1/8	1.0	1000	0.5
	1.6	10	0.001
		1000	0.1
		10000	1

Notes:
1. 10 μl/s slow analytical; 100 μl/s fast or semi-prep; 1000 μl/s prep.
2. Inch–metric conversion of tube sizes are given in section 10.5.
3. Values for pressure drop are well rounded to take account of the wide variations in commercial tube bores and the fourth-power dependence on diameter.

Liquid Velocity in tubing

summary
- **tubing that leads to negligible zone broadening also leads to negligible time lag**

bore mm	thou	area mm^2	volume in 1 m mm^3 or μl	velocity in m/s at flowrate 10 μl/s	100 μl/s	1000 μl/s
0.2	8	0.03	30	0.33	3.3	33
0.3	12	0.07	70	0.14	1.4	14
0.5	20	0.2	200	0.05	0.5	5
0.8	32	0.5	500	0.02	0.02	2
1.6	63	2	2000	0.005	0.05	0.5

12.5 Solvents—properties, problems and purification

summary • **quite few solvents ideal for LC and available commercially, and of suitable quality**

In practice, choice of solvents is mainly governed by what is commercially available, and a plethora of odd solvent brews tends to emerge. In many, but by no means all cases, other solvent combinations would perform just as well. The short list of 'first choice' solvents (section 12.5.1) is intended to make a small contribution to reducing the number of solvents in use: many separations will require others.

The usual reasons for selecting a 'second choice' solvent (section 12.5.2) are:

- higher sample solubility,
- higher solvent solubility or miscibility range,
- acidity or basicity,
- different selectivity makes a particular separation possible.

Desirable properties of LC solvents

summary
- **cost—£4 ($10) per litre maximum with £2 ($4) per litre a target**
- **availability at suitable purity and stability**
- **acceptable toxicity**
- **UV transparency**
- **low viscosity to reduce pressure drop**
- **high diffusivity for sharper peaks**
- **fitting in a range of solvent polarities**
- **low residues for sample recovery without contamination**
- **solvent power**
- **miscible with other solvents to form useful mixtures**

Viscosity

This should be as low as possible to reduce the pressure required to force the liquid through the column, and to ensure a high rate of equilibration (the molecular diffusion coefficient depends mainly on viscosity—see section 12.6) and thus good efficiencies.

Pressure drop is now of less importance because smaller particles mean shorter columns and thus effective pressure requirements have fallen. Nevertheless a low viscosity remains desirable, and values above 1 cP (10^{-3} Ns/m^2) are less attractive.

Solvent strength

According to a naive view of the theory of LSC, a mixture of only two solvents, one of high and one of low solvent strength, will be needed. In practice they may not dissolve in each other over the total required range and, more important, other factors may mean that the required selectivity to give a separation is not achieved. A range of solvent strengths is needed to fit into an eluotropic series. The tabulation in section 12.5.1 is based on the famous list by Snyder[8].

UV transparency

This property is not important chromatographically but vital to most users with UV detectors, and is the greatest source of difficulty. Very small amounts of impurities (or stabilizers) can make a solvent UV opaque, or shift the cutoff to much higher wavelengths. With variable wavelength detectors the requirement for solvents transparent below 220 nm is most difficult to achieve, and some care is needed to avoid contamination of the purified solvent by the pumping system. Gradient elution, use of which is likely to increase as more difficult separations are demanded, also makes even more stringent demands on good UV transparency to avoid violent baseline shifts. It is of course possible to increase the absorbance of one component, to balance eluents: antioxidants have been suggested for this purpose.

Molecular weight

Low molecular weight is desirable to reduce the extent and complexity of mass spectra resulting from residual traces of solvent when peaks are identified by mass spectrometry, a most powerful adjunct to LC. It is simpler to standardise on low molecular weight solvents so that when identification information is required, new separation systems do not have to be developed.

Boiling point

Sample recovery for subsequent analysis by other techniques is easiest if the solvent boils below 100°C. But *too* low boiling solvents cause trouble with reciprocating pumps because they boil instead of being drawn into the pump. Pentane and diethylether are little used because of their low boiling points.

Non-volatile matter

When solutes are to be recovered, for identification or preparation, the residues from solvent after evaporation need considering. Many commercial solvents are distilled from metal plant and contain metallic contamination: iron, for example, can cause problems with NMR. Materials redistilled from glass are better. Particular care is needed to ensure that plasticisers, another almost ubiquitous source of contamination, are eliminated and excluded. A small plastic component, such as an unlined bottle cap, can easily spoil a carefully purified solvent. Suppliers are reluctant to include useful specifications on residues because they rarely have facilities like MS to measure and check residues levels. Users must therefore run blank determinations on all solvents (and batches) used to recover solutes.

Miscibility

A number of attractive polar compounds, mainly acetonitrile and methanol, are not completely miscible with low polarity components, for example hexane. Ethanol or isopropanol have been substituted for the polar component, and cyclohexane, pentane or trimethylpentane for the non-polar.

Ability to dissolve sample

This solvent property is important in adsorption chromatography, particularly for preparative work where the sample solubility may limit the amount that can be injected. In permeation chromatography it is often the main criterion on which a solvent is chosen for so many of the polymers of interest are only reluctantly soluble, even at the higher temperatures which are used to increase solubility and decrease viscosity.

Notes on Purification and Properties of some Solvents[20]

acetonitrile

This solvent has become a firm favourite with many chromatographers. It has good viscosity, bp, UV, solvent properties and low molecular weight, but many feed-stocks are contaminated by aromatics or other compounds used in azeotropic removal of water. Acrylonitrile is usually the main contaminant. The purification of acetonitrile has been described[122]: the authors stress the importance of the process used to make the feedstock. Material of this quality is now available from several specialist suppliers.

methylene chloride

A rather high cut-off (~235 nm) is the main disadvantage of this solvent with good solvent properties, low viscosity and molecular weight. Material comes from many sources and is stabilised with a wide variety of proprietary compounds, often mixtures, to prevent oxidation and hydrolysis. These additives may make the material slightly acidic or basic which may change the chromatographic properties. Material redistilled from glass and stored in the dark under nitrogen, is much more stable and should prove a more reproducible eluent. The types of stabiliser used and their removal have been reviewed[20].

hexane

Aromatics (and other impurities, perhaps olefins below 220 nm) cause problems with an otherwise ideal solvent of low viscosity, good UV transparency, molecular weight and boiling point. Material with suitable specification is now available from several suppliers at prices below the spectroscopic grade prices. The material is usually oleum washed to remove aromatics and this leaves the final product slightly but definitely acid.

tetrahydrofuran

The difficulty with this solvent is that it reacts with oxygen to form peroxides thus usually requiring redistillation just before use. Provided they are in solution, peroxides per se are not dangerous but are a serious hazard if the liquid is distilled without their prior removal. Antioxidants like butylated hydroxytoluenes (BHT) or quinol are usually added to prevent peroxide formation, and this material is usually satisfactory for permeation chromatography with RI detection unless solutes are to be recovered. However this stabiliser makes the material useless for most applications where UV or IR detection is to be used. Bottling pure material under nitrogen in the dark can prevent most peroxide formation and this material can be used in the UV down to about 220 nm. But if oxygen is allowed to react with this unstabilised material peroxides are formed and eventually a substantial fraction of the material will be oxidised to n-propyl formate and become UV opaque to above 300 nm. Unstabilised material has a limited life once the bottle is opened and must not be distilled without peroxide removal.

chloroform

Analar material (and the BP and USP XVI grades) are quite pure but stabilised with 0.5–2% of ethanol which prevents phosgene from photo-oxidation building up. Provided the level of ethanol is known to be constant it may be tolerable but in many applications it is removed and a known and smaller amount added. This tiresome procedure may become unnecessary as glass distilled material with a small and precisely controlled ethanol content becomes available. A level of 0.5% controlled to ±5% seems a reasonable target.

o-dichlorobenzene

This solvent is widely used at 130°C for permeation chromatography. The addition of anti-oxidants is recommended. Suitable commercial types are Topanol OC (ICI), Ionol (Shell), Sanatox (Monsanto) or 2,6-ditertbutyl-4-methyl phenol.

trichloroethylene

Like dichloromethane commercial grades are all stabilised with proprietary mixtures (a mixture of additives is chosen for synergistic effect). Amines are commonly used. Purer grades can be kept under nitrogen and in the dark for much longer periods without degradation.

Some suppliers of Solvents specially for LC

Europe:
BDH Chemicals Ltd, Broom Road, Poole, Dorset BH12 4NN
Fisons Scientific Apparatus, Bishop Meadow Road, Loughborough, Leics, LE11 0RG
Fluka AG Chemische Fabrik, CH-9470 Buchs, Schweiz
Hopkins and Williams, Chadwell Heath, Essex
E. Merck AG, 61 Darmstadt, Bundesrepublik, Deutschland
Rathburn Chemicals Ltd, Caberston Road, Walkerburn, Peeblesshire, Scotland

USA:
J.T. Baker Chemical Co., 222 Red School Lane, Phillipsburg, N.J. 08865
Burdick and Jackson, Muskegon, Michegan, USA
Eastman-Kodak, Eastman Organic Chemicals, Rochester, N.Y. 14650
Mallinckrodt, Science Products Division, PO Box 5439, St. Louis, Missouri, USA 63147
MC/B, 2909 Highland Avenue, Norwood, Ohio 45212

Not all grades for chromatography are satisfactory for all applications. Spectroscopic grades may be suitable but are often much more expensive, reflecting small consumption rather than necessarily higher quality than equivalent LC grades.

12.5.1 Solvents – See inside front cover

12.5.2 Most Useful Solvents for LC Using Variable Wavelength UV Detectors

Solvent	Eluting[a] Power (s^o Al_2O_3)	Viscosity N/s x 10^3 20°C	Diffusion[b] $m^2/s \times 10^9$ 20°C	UV[c] cutoff nm	Comments
pentane	0.00	0.235	4.9	210	Very volatile.
hexane	0.01	0.326	3.8	200	
diethyl ether	0.38	0.23	4.9	220	too low boiling for some pumps
chloroform	0.40	0.58	2.5	245	uv limit rather high
dichloro-methane	0.42	0.43	2.8	245	uv limit rather high
tetrahydro furan	0.45	0.55	2.0	220	stabilizer free grade for good uv causes peroxide problems.
dichloro-ethane	0.49	0.79	1.6	230	more viscous but better uv than CH_2Cl_2 or $CHCl_3$
aceto-nitrile	0.65	0.37	2.4	210	Chromato-graphically good.
methanol	0.95	0.60	1.7	210	rather low misci-bility range hexane and CH_3CN
water	large	1.00	0.9	200	rather viscous and causes problems with dissolved gas in detectors.

a s^o polarity values from L.R. Snyder[19].

b Diffusion coefficients calculated from the Wilke-Chang equation, or taken from their data using nitrobenzene or toluene (which have similar molar volumes) as solute. The accuracy of this equation is open to some doubt[9] and some values may be in error. In general an inverse relationship with viscosity is found. see section 12.6

c UV cutoff values are approximate values for selected LC grades available at reasonable cost giving 10% transmittance (1 absorbance unit).

Some dense solvents for balanced density slurry packing

compound		density (g/ml)
carbon tetrachloride	CCl_4	1.59
trichloroethylene	$CHCl\,CCl_2$	1.46
1,1,2,2-tetrachloroethane	$CHCl_2CHCl_2$	1.59
pentachloroethane	$CHCl_2CCl_3$	1.68
1,2-dibromoethane	CH_2BrCH_2Br	2.18
1,1,2,2-tetrabromoethane	$CHBr_2CHBr_2$	2.96
trichlorotrifluoroethane	CCl_3CF_3	1.52
iodomethane	CH_3I	2.28
tetrachloroethylene (per)	CCl_2CCl_2	1.62
bromoform	$CHBr_3$	2.89

Viscosity of aqueous–organic mixtures

Aqueous–organic mixtures have unexpectedly high viscosities, and presumably the diffusivity of solutes is correspondingly slower. At some composition the viscosity is higher than for either water or the organic component, which is usually less viscous than water.

organic component	viscosity mN s/m^2	maximum viscosity of aqueous–organic mN s/m^2	composition % organic
formic acid		1.5	70
acetic acid	1.2	2.7	80
ethanol	1.5	2.8	40
methanol	0.59	1.8	40
acetone	0.32	1.2	10
acetonitrile	0.38	1.1	10

(from reference 79, section D218, except acetonitrile[372])

As Guggenheim has remarked *"any theory of liquids that applies to water is wrong".*

12.6 Diffusion Coefficients

summary

- **sparse experimental data**
- **empirical equations give variable accuracy**
- **in practice select least viscous solvent**
- **hexane, acetonitrile, methanol and water may be preferable for detailed column assessment**

The problem of obtaining practical values of diffusivity for chromatography (and many other chemical engineering problems) has been tackled by many. Since the diffusivity is different for each solute in each different solvent there is a need for a predictive equation. None of the many equations developed are entirely satisfactory either as a theoretical model or in the accuracy of predicted data. Since entirely theoretical equations such as that due to Einstein[51] (the Stokes-Einstein equation which predicts diffusivity proportional to $T/\eta V^{0.33}$ where V is the molar volume) do not give practical accuracy, most work has centred on empirical equations to extrapolate from the inevitably sparse experimental data. Gambill[53] gives a useful summary of the problem.

The Wilke-Chang equation[45] is most widely quoted

$$D_m = 7.4 \times 10^{-12} \sqrt{\Psi M_{solvent}}\, T/\eta\, V_{solute}^{0.6}$$

where Ψ is a solvent association factor (1 for non-polar solvents, 1.9 for methanol, 1.5 for ethanol, and 2.6 for water, $M_{solvent}$ the molecular weight (g), T the temperature (K), and V_{solute} the molar volume (ml) that is molecular weight/density. D_m has units of m^2/s.

Other workers disagree, in part with the benefit of more data than that available and determined by Wilke and Chang. (Experimental determination is rather tedious – Chang and Wilke[49] describe a typical example.) The most frankly empirical equation is by Hayduck and Cheng[50] who propose only correlation with solvent viscosity η. Their equation is $D = A\eta^B$ where A and B are solute parameters. This obscures the disputed importance of molar volume and shape (Stokes-Einstein predicts $V^{0.33}$ but Wilke and Chang find $V^{0.6}$ fits the data better). De Kee and Laudie[48] give a nomogram for quick calculation from this equation but data for few chromatographically interesting solute-solvent pairs is available. It is clear too that many, if not most, chromatographic separations involve solvent-solute interactions that make experimental determination essential for reliable elucidation of the effect of diffusivity.

In practical solvent selection there is little problem—unless there are other overriding criteria (and there often are)—select the least viscous solvent. For work involving assessment of column performance comparison between different laboratories is likely to be best served by the use of the same solvents (and solutes too where applicable) thus reducing the number of possible variables. Fortunately the effect of diffusivity is small but differences as large as a factor of three, for example between hexane and cyclohexane, can hardly be ignored. Elimination of this variable rather than correction seems preferable.

Some data calculated from this equation are given below:

solvent	viscosity Cp or m N s/m^2	temperature C	solute	molar volume m^3 x 10^6	diffusivity m^2/s x 10^9
pentane	0.235	20	toluene	106	4.8
hexane	0.326	20	"	"	3.8
"	0.294	25	"	"	4.2
cyclohexane	1.0	20	"	"	1.2
chloroform	0.58	"	"	"	2.5
diethyl ether	0.23	"	"	"	4.9
tetrahydrofuran	0.55	"	"	"	2.0
dichloroethane	0.79	"	"	"	1.6
acetonitrile	0.35	"	"	"	2.4
methanol	0.60	"	"	"	1.7
"	0.40	50	"	"	2.8
ethanol	1.2	20	"	"	0.91
water	1.00	"	"	"	0.90
"	0.547	50	"	"	1.82
"	0.282	100	"	"	4.1
hexane	0.326	20	"	"	3.8
"	"	"	benzene	89	4.2
			bromobenzene	104	3.8
"	"	"	nitrobenzene	102	3.8
"	"	"	naphthalene	136	3.2
"	"	"	phenanthrene	173	2.8
"	"	"	acetophenone	117	3.5
methanol/H_2O 40/60 v/v	1.5	"	phenol		0.74
			acetone		0.82
					0.64
			2, 6-xylenol		0.86
70/30 v/v	1.48	"	acetone		0.78
			phenol		0.70
			o-cresol		

Solvents for Chromatographic Research

Perhaps a simplified method for determining diffusion coefficients[87] described recently will encourage a few more experimental determinations.

One test system has been proposed[52] and many others will be needed to test other chromatographic factors. From the long list of useful solvents (discussed further in section 12.5) a few seem most useful for this type of study.

Hexane seems most useful from the several possible non-polar solvents. It is now available freely at suitable purity and cost, a problem in the past, and has a low viscosity. Of the other contenders, pentane might be preferred for its lower viscosity but is less widely available and more expensive for LC grades and, more important, is too low boiling for some types of pump. Higher alkanes and cyclohexane have no advantages over hexane and all have the disadvantage of higher viscosity. Mixtures containing small amounts of more polar components will have similar diffusivity to that of hexane or other major component alone.

Acetonitrile must be the solvent of choice for medium polarity now that it is available in suitable LC grades, with *methanol* and *water* completing the spectrum. If the gap in eluting power between hexane and acetonitrile is considered too great one might include an ether solvent: either diethyl ether, which, like pentane, is difficult to use with some pumps, or tetrahydrofuran, but both cause problems with peroxides.

12.7 Compressibility

summary • **1% volume change at 100 bar (1500 psi)**

Liquids are hardly compressible. Typical values for common LC solvents are tabulated below[31]. Compressibility is the fraction volume change resulting from a pressure change.

Liquid	Compressibility m^2/N or Pa^{-1} $\times 10^9$
n-pentane	3.14
n-hexane	1.61
chloroform	0.97
diethyl ether	1.87
methanol	1.23
ethanol	1.11
water	0.46

Thus if the typical compressibility is taken as $\sim 10^{-9}$ m^2/N then an applied pressure of 100 bar or $10^7 N/m^2$ results in fractional volume change of 10^{-2} or 1%. Alternatively each bar pressure changes the volume by 10^{-4}.

12.8 Particle Size Measurement

summary
- **microscopic examination quick and informative**
- **precise measurement and definition difficult especially for wide range of size**
- **comparison of different shaped particles or with different method of measurement needs care**

Measurement of particle sizes of small LC packings is not simple[19,21]. Description of the particle size distribution is even more difficult. The porosity of most LC materials is an added complication.

Allen[21] reviews critically the many methods and problems in particle size measurement. He also describes the methods of size classification which might be used to prepare LC packings.

Sieving

Sieving gives satisfactory results above about 100 μm but becomes less satisfactory for smaller particles. Various methods of agitation are used to persuade all those particles that are in fact small enough to pass through the sieve. Commercial equipment aims to break up aggregates and prevent blinding the sieve. Wet sieving is one method of achieving this. But for microparticles under 20 μm diameter, sieves are little help (see also section 10.7).

Sedimentation

Comparisons between spherical and granular or otherwise irregular packings are fraught with difficulty. The 'equivalent spherical diameter' concept comes from particle size assessment by sedimentation. Stokes Law relates the settling rate to the diameter of a falling sphere.

$$v = (\rho_s - \rho_F)\, gd^2/18\eta$$

where
ρ_s = solid density
ρ_F = fluid density
g = acceleration due to gravity
η = viscosity of fluid
d = diameter

Non-spherical particles falling at the same rate as spherical spheres are said to have an equal 'equivalent spherical diameter'.

There are a number of methods, and commercial devices, of measuring the rate of sedimentation. All become less reliable for few micrometer sizes. In particular the results may be meaningful and quite precise on a particular instrument but not accurate absolutely and thus not strictly comparable with results obtained on another instrument. Specification of the method of size determination is therefore essential. Two problems are apparent: finding the real density of the particle and ensuring that none stick together, when they would sink much faster.

For small particles air can be used as the fluid with a settling distance of about a metre. Settling rates in liquid are too slow to measure directly except on an instrument such as the SediGraph 5000 made by Micromeritics Instrument Co which moves the sensing beam of soft x-rays upwards instead of waiting for the particles to fall into the beam. This reduces the analysis time considerably for very small particles and gives a direct plot of size against cumulative mass percent finer.

The density can be calculated from the measured pore volume and matrix density (2.3 for silica and 3.5 for alumina) or by direct measurement of the density of a slurry of known concentration in water. It is usual to add a surfactant to disperse aggregates and to agitate in an ultrasonic bath and/or hold the slurry under vacuum to remove all air from the pores and ensure they are full of liquid. Air in the pores would lead to serious errors of course. Typical silicas have densities of 0.8–1 g/ml.

Coulter Counter

The Coulter technique is a method of determining the number and size of particles suspended in an electrolyte by causing them to pass through a small orifice which is immersed in electrode. The changes in resistance as particles pass through the orifice generate pulses whose size is related to the volume of the particle.

Results for particles <1 μm provide a useful basis for comparing the efficiency of removing fines, but the results are not easy to correlate with results from other methods of particle size measurement[21].

The Coulter counter is a useful device for measuring numbers of particles within a given volume range. However there are some problems in relating results with those obtained by direct measurement under the microscope or sieve, or sedimentation. This is because the pores do not contribute fully to the measured diameter. A correct value for the density is needed to compensate for the pore volume. Calibration against closely sized spheres could also give precise comparison with other methods. Particle shape also has some effect. Long thin fragments give results which differ most from that expected from their volume.

Nevertheless this technique can give useful results over the range from 0.2 μm to 20 μm, which is the diameter range of most current interest.

Microscopy

Measurement from photomicrographs, or even direct from the microscope, is most useful because it is an absolute method. A quick assessment by eye (because it is backed by a powerful data processor) can give a useful indication of particle size distribution and, for similar distributions at least, a good sensitivity to changes in mean particle size. It is always recommended as a way of detecting gross errors in results from other methods. Unfortunately it is impractical to measure enough particles to get a statistically significant result for much more than the mean (and even that is tedious) without some form of automation. Such image analysing computer systems now exist but a detailed description of the size distribution based on this type of instrument has yet to be published. It is also difficult to assess visually the size of irregular particles so this method is most appropriate for spherical particles.

More than the other methods of determining particle size, microscopy presents a major sampling problem. Because relatively few particles can be measured compared to the other techniques, it is difficult to get a representative sample onto a microscopic slide. One method that has been used is to dip a microscope slide coated with a very thin film of grease into the bulk sample: a reasonably representative monolayer will adhere.

Even with closely sized spheres it is not easy to measure mean particle size to better than ±10%. Thus apparently identical batches may produce plate heights which differ by 20%. This variation, which is important only in detailed studies of column materials and packing methods, can be eliminated by using one homogeneous batch. With irregular particles, or wherever the range of size is great, the problem is much worse in that to calculate reduced parameters some mean particle size must be taken for d_p. The term 'mean particle size' is widely used; the definition in use is more rarely given; and as the range of particle sizes increases its meaningfulness declines. Comparisons based on mean particle size need to be made with care. Variations in reduced plate height of ±30% may not be significant if different manufacturers nominal values are used as the basis for comparison. Note too that volume or mass distribution comes naturally from sedimentation results; number distribution by contrast results naturally from microscopy. Expressing results as a number percent weights the fine end of the spectrum and appears to reduce the 'mean size'.

It is hardly surprising therefore that most detailed theoretical studies have used rather large spherical materials.

Permeability

summary • **chromatographically the most useful estimate of mean size**

Since to the chromatographer the particle size is of fairly academic interest, useful mainly for calculation of reduced parameters, it is not unreasonable to infer particle size from the observed specific column permeability using the equation

$$d_p = \sqrt{\phi'/K^\circ} = \sqrt{\phi' \eta L^2/Pt_o}$$

The difficulty with this approach is in choosing a value for ϕ' which must itself depend on the packing method, and other factors. At present a value of 500 appears to be most appropriate[204] for porous particles which have been slurry packed.

Values obtained by this method are particularly useful when considering materials with a wide particle size range. Fines can cause a comminuted material to have a particle size estimated from impedance to be much less than that predicted by other methods.

12.9 Hazards and Safety in LC

summary
- **high pressure hazard small**
- **fire main danger and requires precautions, especially for preparative work**
- **toxicity hazard small but operators should wear safety glasses**

The description 'high pressure' arouses interest from safety officers. There are few grounds for concern: LC is no more dangerous than most other analytical methods. Nevertheless attention to safety requirements is important at a few critical points, and prudent at some others. This section attempts to list the main dangers and suggest ways to avoid accidents. (An accident has been defined as the meeting of the unprepared and the unexpected[75].) The main hazards arise from high pressure, fire and explosion, and toxicity of the chemicals used[74].

High pressure

The danger from high pressure is slight. This is because liquid is barely compressible and the energy stored, or available for release on fracture, is small. Hydraulic pressure ratings and safety margins should be applied. Gas pressure needs to be treated with far more respect for the stored energy can fire fragments at high velocity causing serious injury. Excessive liquid pressure rarely leads to fracture—a component will start to leak before breaking. Even glass tube can be broken without any excitement at 100 bar liquid pressure. Pressure limiting devices are therefore mainly to protect the equipment from damaging itself rather than for the benefit of the operator. Methods are discussed in detail in section 7.3 on pumps. In summary, constant pressure pumps should be controlled by limiting the applied gas pressure, whether air or nitrogen, and also the maximum flow limited by selecting a suitable pipe bore; constant flow pumps need some pressure limiting devices to vent excess pressure or switch off the pump. Direct gas-driven pumps should not be used with glass columns.

Fire or Explosion

This is the most serious hazard: sensible precautions can minimize but not eliminate it. The following methods are worth considering:

- **Removing sources of ignition.** Smoking is obviously hazardous. Pneumatic actuation of valves and pumps can take electrical devices out of range of the bulk of solvent. Inevitably detectors are electrical and it is impractical at present to make them suitable for use in classified zones where solvent vapour may occur. Flame-proof motors prevent sparks from the motor brushes igniting vapour outside the motor. (This is achieved by encasing the motor in a metal case strong enough to withstand an internal explosion and with narrow gaps so that flames cannot propagate the ignition outside the case.) Flameproof motors become rather large as the power rating rises; pneumatic operation may be more convenient. Non-sparking switches, for example reed relays and mercury switches, reduce risk but are less effective than removing all electrical equipment because loose connections can still cause sparks.

• **Removing oxygen by nitrogen purging.** This is particularly desirable if there is any danger of explosive vapour building up inside the apparatus, usually in the oven. Since most LC solvents form explosive mixtures above 1% (v/v), and 1 mmol ($\approx$0.1g) gives a 22.4 ml of vapour at ambient conditions, even this small weight will give an explosive mixture in a 2 litre oven. Several litres of nitrogen are needed to purge this effectively even if flow paths are well designed. On safety grounds (as well as cost) column temperature control by passing thermostatted water through jackets must be preferred. Solvent reservoirs may also benefit from purging.

• **Limiting solvent.** As little solvent as possible should be outside fireproof storage containers such as metal lined cupboards or closets. If fire does start it is important to contain it. A simple way, particularly desirable for preparative work where larger amounts of solvent are inevitable, is to place equipment in metal trays whose volume is enough to contain any spillage from the reservoirs. It may be prudent to place the detector in a separate tray. Any fire will be localised to the tray, more easily fought with an extinguisher (which should be close at hand), and less likely to ignite the fabric of the building or the bench.

Toxicity

Most 8 hr exposure limits for useful LC solvents are above 100 ppm but a few are nearer to 10 ppm. (Many levels are currently being revised and will therefore not be listed here.) With adequate general laboratory ventilation, few LC solvents are so toxic that long-term exposure levels need be exceeded. An exception is the gas vented from direct gas-driven pumps such as coil pumps: this gas needs to be vented into a fume duct.

It is always sensible and easy (especially with rather toxic solvents like chloroform and acetonitrile) to minimize solvent evaporation from reservoirs and collection vessels. Simple precautions like not collecting solvent in open containers can reduce evaporation substantially, and avoid the need for draconian safety measures.

summary • **wear safety spectacles when using LC**

Acute toxicity hazards can arise from leaks. The main danger is that a jet of liquid will get into the operator's eye. Safety glasses render this risk acceptably low: fortunately most solvent mixtures are relatively innocuous and serious eye injury is unlikely provided this precaution is taken.

12.10 Column Parameter Calculation Program

summary • **most parameters of a column can be calculated from 10 measurements**

Evaluation of columns used in liquid chromatography has been described in section 8.8. Calculation of the parameters from raw data is conveniently accomplished using the computer program in BASIC given below. Ten measured parameters are used:

Line 1: Packing type or identification code; length (mm) to 1 mm; column internal diameter (mm) to 0.1 mm; particle diameter (μm).

Line 2: 'Solvent', eluent or unretained peak time (s) to 0.5 s; retained peak time (s) to 0.5 s; width at ½ height of retained peak (s) to 0.2 s. Can also be entered as mm for other chart speeds.

Line 3: Pressure (bar) (to 2%) (multiply value in psi by 0.069); flow rate (ml/s) (volume (ml) collected in time (s)); injection volume (μl).

To run the Program type RUN (c/r).

(c/r ≡ RETURN ≡ carriage return which is typed after each reply). If the response to the question 'instructions?' does not start with N, full instructions are printed. Otherwise data is requested immediately immediately. After each report further data is requested. Program execution is terminated by replying 'stop' to the question 'SAME COLUMN?'

If the values for viscosity and diffusion coefficient are not suitable then different values should be substituted on the appropriate program line. Sections 12.5 and 12.6 give some values. Viscosity and diffusivity values should obviously be those applying at the temperature used. Different chart speeds can be used if the value of C in the program is changed.

Notes on potential causes of error:

- The actual bed length and internal diameter and particle diameters are often quite different from the nominal sizes.
- The baseline crossover point is the best estimate of the mean solvent velocity from a derivative peak.
- Width at half-height is recommended for assessment of peak width because it is convenient, unequivocal and not subject to artistic licence. Measurement should be from centres of lines: errors in this measurement can produce large errors in the derived performance parameters. Measurement to 0.1 mm is just feasible: thus a minimum half height width of 2 mm is required to achieve a ±5% precision, and hence ±10% plate number precision.
- Pressure drop across injection ports, column end frits, pipework and detectors may not be negligible. Flow rates should be measured at the column exit with the detector disconnected. Nominal pump flow rate settings are not usually accurate especially at high pressures.

```
100 REM COLUMN PARAMETER CALCULATION PROGRAM
110 REM VER 1.7 NOV 76
120 REM AUTHOR PAUL BRISTOW, QUALITY CONTROL DEPT, ICI PHARMACEUTICALS
130 REM DIVISION, HURDSFIELD INDUSTRIAL ESTATE, MACCLESFIELD, CHESHIRE
140 REM RUN ON IBM CALL SYSTEM
150 PRINT 'INSTRUCTIONS';
160 INPUT A$
170 IF STR(A$,1,1)='N' GOTO 430
180 PRINT'DATA REQUESTED BY ? ON 3 LINES,'
190 PRINT'ENTER EACH VALUE SEPARATED BY COMMAS, AND IN UNITS GIVEN.'
200 PRINT'CHART SPEED 1 MM/SEC IS MOST CONVENIENT.'
210 PRINT'AFTER EACH REPORT FURTHER DATA SETS ARE REQUESTED.'
220 PRINT'TO STOP ENTERING DATA REPLY "STOP".'
230 PRINT'IF YOU MIS-TYPE, RETYPE ONLY THE WRONG LINE.'
240 PRINT'TYPE IN :-'
250 PRINT'PACKING TYPE, LENGTH (MM), BORE (MM), PARTICLE DIA (UM), RETURN'
260 PRINT'2ND LINE: UNADSORBED PEAK (MM), PEAK (MM), ';
270 PRINT'PEAK 1/2 HT WIDTH (MM)'
280 PRINT'3RD LINE: PRESSURE (BAR),FLOWRATE (ML/S), INJ VOL (UL)'
290 PRINT 'FOR SAME COLUMN, INPUT 2ND AND 3RD LINES ONLY.'
300 REM C=CHART SPEED IN MM/SEC, 1 ASSUMED; ALTER C FOR OTHER SPEEDS
310 REM SPEED SEC/CM     MM/MIN     C MM/SEC
320 REM        1           600         10
330 REM        5           120         2
340 REM       20            30         0.5
350 REM      60             10         0.1667
360 REM     120              5         0.0834
370 REM                      2         0.0333
380 REM                      1         0.01667
390 REM       60        600/HR         0.1667
400 REM      300           120         0.0333
410 PRINT 'HEXANE ELUENT AND NITROBENZENE SOLUTE AT AMBIENT ASSUMED,'
420 PRINT 'ALTER VISCOSITY AND DIFFUSION COEFFICIENT IF NECESSARY.'
430 V=0.326E-3
440 REM VISCOSITY NSEC/M2
450 F=3.8E-9
460 REM DIFFUSION COEFFICIENT M2/SEC
470 REM DIFFUSION COEFFICICIENT CALC FROM WILKE-CHANG FOR NITROBENZENE
480 INPUT A$,L,I,D
490 INPUT T0,T,W
500 INPUT P,Q,J
510 C=1
520 T0=T0/C
530 T=T/C
540 W=W/C
550 PRINT
560 PRINT'COLUMN PARAMETERS     VER 1.7      TYPE ';A$
570 PRINT
580 PRINT
590 PRINTUSING 1140,L,I,D
600:COLUMN TUBE VOLUME ###.## ML     AREA ### MM2
610 PRINTUSING 600,0.785E-3*I*I*L,0.785*I*I
620 D1=I*I*1E3/(L*D)
630:KNOX-PARCHER RATIO ####.#
640:KNOX-PARCHER RATIO ####.#       INFINITE DIAMETER MODE POSSIBLE
650 IF((I-2)*(I-2)/(L*D*1E-3))<(2.4+32/(1E-9*L*D/(T0*F))) THEN 680
660 PRINTUSING 640,D1
670 GOTO 690
680 PRINTUSING 630,D1
690 Q=Q*1000
700 REM Q FLOW UL/SEC
710 R=T/W
720 N=5.5452*R*R
730 REM A B LITTLEWOOD, GAS CHROMATOGRAPHY, ACAD PRESS (1970), P 154
740 REM 5.5452=8 LN(2)
750 H=0.001*L/N
```

```
760 PRINT
770 PRINTUSING 1210,R,N,H*1E6
780 PRINTUSING 1300,126.48*SQR(H*L),SQR(20*F*1E6*T0+L*D*1E-3+1.6*J**0.67
),J
790 REM ASSUMES ON-COLUMN INJECTION SLUG SPREADING TO A SPHERE
800 PRINTUSING 1240, T0,T,W*1.70
810 PRINTUSING 1200,O*T0/1000,1.70*O*T0/R,1.70*W*O
820 PRINTUSING 1270,O,L/T0,P
830 I=I*1E-3
840 L=L*1E-3
850 P=P*1E5
860 D=D*1E-6
870 REM NOW IN SI UNITS
880 PRINT
890 PRINTUSING 900,V,F*1E6
900:ELUENT VISCOSITY #.#!!!! N S/M2    SOLUTE DIFFUSIVITY #.#!!!! MM2/S
910 PRINTUSING 1250,H/D,L*D/(T0*F),0.001*L/D
920 F0=P*D*D*T0/(V*L*L)
930 K=L*L*V/(T0*P)
940 Z=500
950 PRINT
960PRINTUSING 1230,1E6*SQR(Z*V*L*L/(P*T0)),1E3*O*V*L/(P*0.785*I*I)
970 PRINTUSING 1260,F0,K*1E12
980 IF F0<3000 THEN 1000
990 PRINT "ANOMALOUS 0'?"
1000 V1=0.785*I*I*L
1010 V2=O*T0*1E-9
1020 P0=V2/V1
1030 R2=(1-T0/T)*(1-T0/T)
1040 REM J H PURNELL & C P QUINN, GAS CHROMATOG, ACAD PRESS (1960) P 184
1050 PRINTUSING 1280,0.01*H*H/K,P0
1060 REM J H KNOX & M SALEEM, J CHROM SCI, 7, 614 (1969) EQ 21 (& 14)
1070 IF P0>0.4 GOTO 1090
1080 PRINT 'UNLIKELY POROSITY'
1090 IF P0<1 GOTO 1110
1100 PRINT 'IMPOSSIBLE POROSITY - ELUENT TIME, DIMENSION OR FLOW ERROR?'
1110 PRINT
1120 PRINTUSING 1220,(T-T0)/T0,T0/T
1130 PRINTUSING 1290,N*R2,N*R2/T
1140:LENGTH #### MM  INTERNAL DIA ##.# MM  PARTICLE DIA ##.# UM
1150 K1=10
1160 REM K1 IS ARBITRARY CHOICE
1170 PRINTUSING 1310,1+0.6*SQR(N)*LGT(1+K1),K1
1180 REM GIDDINGS ANAL CHEM 39,1027 (1967)
1190 REM J.N.DONE & J.H.KNOX, J.CHROM. SCI., 11,606 (1972)
1200:ELUENT VOL ###.## ML  ELUENT PEAK VOL #### UL  PEAK VOL ##### UL
1210:DISPERSION RATIO ###.#  PLATE NO ######  PLATE HT ####.# UM
1220:PEAK K' ##.##      RETENTION RATIO R #.###
1230:EFFECTIVE SPHERICAL DIA ###.# UM   SPECIFIC PERM ##.#### UM2
1240:ELUENT TIME #### S    PEAK TIME #### S    WIDTH ###.# S
1250:REDUCED H ##.#  REDUCED VELOCITY ###.#  REDUCED LENGTH ####000
1260:FLOW RESISTANCE 0 ####      PERMEABILITY ##.##### UM2
1270:FLOWRATE #### UL/S    VELOCITY ##.## MM/S    PRESSURE ###.# BAR
1280:SEPARATION IMPEDANCE ###00         POROSITY .##
1290:SEPARATION NO ######    EFFECTIVE PLATES/S ### FOR PEAK
1300:PEAK DISPERSION  AXIAL ###.# MM  RADIAL ##.# MM    INJ VOL ##### UL
1310:PEAK CAPACITY #### FOR K' = ##
1320 PRINT
1330 PRINT
1340 PRINT 'SAME COLUMN';
1350 INPUT A$
1360 IF A$='STOP' GOTO 1440
1370 IF STR(A$,1,1)<>'Y' GOTO 150
1380 INPUT T0,T,W
1390 INPUT P,O,J
1400 I=I*1E3
1410 L=L*1E3
1420 D=D*1E6
1430 GOTO 690
1440 END
```

```
RUN

INSTRUCTIONS?Y
DATA REQUESTED BY ? ON 3 LINES,
ENTER EACH VALUE SEPARATED BY COMMAS, AND IN UNITS GIVEN.
CHART SPEED 1 MM/SEC IS MOST CONVENIENT.
AFTER EACH REPORT FURTHER DATA SETS ARE REQUESTED.
TO STOP ENTERING DATA REPLY "STOP".
IF YOU MIS-TYPE, RETYPE ONLY THE WRONG LINE.
TYPE IN :-
PACKING TYPE, LENGTH (MM), BORE (MM), PARTICLE DIA (UM), RETURN
2ND LINE: UNADSORBED PEAK (MM), PEAK (MM), PEAK 1/2 HT WIDTH (MM)
3RD LINE: PRESSURE (BAR),FLOWRATE (ML/S), INJ VOL (UL)
FOR SAME COLUMN, INPUT 2ND AND 3RD LINES ONLY.
HEXANE ELUENT AND NITROBENZENE SOLUTE AT AMBIENT ASSUMED,
ALTER VISCOSITY AND DIFFUSION COEFFICIENT IF NECESSARY.
?SPHERISORB S5W,196,8,5
?47.5,114,2.0
< 48.5,0.143,5

COLUMN PARAMETERS     VER 1.7      TYPE SPHERISORB S5W


LENGTH  196 MM  INTERNAL DIA  8.0 MM  PARTICLE DIA  5.0 UM
COLUMN TUBE VOLUME    9.85 ML    AREA  50 MM2
KNOX-PARCHER RATIO    65.3    INFINITE DIAMETER MODE POSSIBLE

DISPERSION RATIO  57.0  PLATE NO  18016  PLATE HT   10.9 UM
PEAK DISPERSION  AXIAL   5.8 MM  RADIAL  3.0 MM    INJ VOL     5 UL
ELUENT TIME    47 S    PEAK TIME  114 S    WIDTH   3.4 S
ELUENT VOL   6.79 ML  ELUENT PEAK VOL  202 UL  PEAK VOL   486 UL
FLOWRATE  143 UL/S    VELOCITY  4.13 MM/S    PRESSURE  48.5 BAR

ELUENT VISCOSITY 3.3E-04 N S/M2    SOLUTE DIFFUSIVITY 3.8E-03 MM2/S
REDUCED H  2.2  REDUCED VELOCITY    5.4  REDUCED LENGTH   39000

EFFECTIVE SPHERICAL DIA   5.2 UM   SPECIFIC PERM  0.0375 UM2
FLOW RESISTANCE 0  459     PERMEABILITY  0.05436 UM2
SEPARATION IMPEDANCE  2100     POROSITY .69

PEAK K'  1.40    RETENTION RATIO R  .417
SEPARATION NO    6130    EFFECTIVE PLATES/S  53 FOR PEAK
PEAK CAPACITY    84 FOR K' = 10

SAME COLUMN?STOP
```

12.11 Thin-layer Liquid Chromatography
(compared to Column Chromatography)

summary
- **closely related in theory**
- **put to work in different practical ways**
- **complementary rather than competitive**

TLC plates consist of flat layers of slurry-packed microparticle silica. (HPLC silica makes excellent but expensive and delicate TLC plates.)

Not surprisingly they work very much like thin flat columns. The 'pump' is capillary action drawing the 'eluent' up to 'develop' the plate. Much of the same theory can be applied to this system. Just as the theory of HPLC has predicted the advantages of smaller particles, so does it predict some advantages for smaller particles for TLC, and this prediction has been proven by the results from high performance TLC. Just like HPLC, the engineering problems caused by the small injection and detection systems now determine the ultimate performance.

Historically, TLC predates column chromatography by over a decade with the first recorded application in 1889 by the Dutch biologist Beyerinck. J G Kirchner[382] has reviewed progress made with the technique since then. For more detail one may explore the treatise of E Stahl[9].

Although often presented as competitive techniques, TLC and HPLC are largely complementary. The use of TLC to survey before starting work with HPLC is discussed in section 8.1 where some differences, and problems translating from one technique to another, are noted.

The two main differences are:

- In HPLC the column is usually at, or near, equilibrium with the eluent, whereas for TLC it is normal for the solvent composition to vary up the plate, with distance and time.

- All solutes are visible, if they can be detected, on a TLC plate, whereas only those that have been eluted can be detected by HPLC.

Some simple calculations can show that reduced velocities used in TLC are rather less than in HPLC. This means that the efficiency is a little less than at the minimum of the plate height—velocity curve, and that obstructive factors described by the B-term of the plate height equation contribute most to the observed plate height (at $\nu = 1$, about three-quarters). The quality of packing (A-term) and diffusion or mass transfer rates in the pores (C-term) are much less important than at the higher speeds used in HSLC.

[The importance of the B-term probably explains the round spot shape typical in TLC, compared to the elongated spot characteristic of column chromatography, for example as shown in ref. 3, p. 415. At low reduced velocities the sideways (radial) dispersion would seem to be more important than the front-to-back (axial) dispersion[120].]

Advantages of TLC:

- all solutes are visible (if they can be detected)
- universal detection feasible with suitable spray reagent
- selective detection by choice of reagent, or radiolabelling, or wavelength of scanner
- many samples can be run together on one plate
- natural gradient of elution strength on the plate
- vast literature of developed methods
- equipment is cheap and reliable

disadvantages:

- analysis time for one plate is quite long and cannot be speeded up
- detection sensitivity is often lower than feasible with HPLC
- only a few compounds (<50) can be separated on each plate
- limited range of concentration can be measured (1:100)
- limited resolving power – only a few hundred plates are typical
- multiple development is slow and inconvenient
- accuracy usually ± 20% and at best ± 2%
- quantitation of many spots is slow and expensive to automate

Tasks for which TLC is good:

- as a survey for HPLC
- qualitative analysis of unknown samples
- limit tests for impurities (if a twofold variation is acceptable)
- trying out solvent systems (especially the microcircular system of Stahl ref. 9 p. 73)
- qualitative analysis of many samples simultaneously

and those for which HPLC is better:

- precise and accurate composition measurement (± 0.1 – 1%)
- high speed analysis (~2 min)
- trace analysis
- mixtures with a wide range of component concentrations ($1:10^4$)
- mixtures with very many (up to many hundreds) of components
- where verv high resolving power (10^5 plates) are essential for separation
- for separation by molecular size (permeation chromatography)

INDEX

11.5 Equations

	Theoretical	Practical
plate no	$N = 5.54\,(t_r/w_{½ht})^2$ $= 16\,(tr/w_{base})^2$ $= L/H = l/h$	as theoretical. tr and w must be in the *same* units—time, volume, distance, or chart distance
width at base (4σ)	$w = 1.70\ w\,½ht$ $= 1.20\ w\,¼ht$	any units
HETP	$H = L/N = \sigma_L^2/L$	$H\ (mm) = L\ (mm)/N$
reduced plate ht	$h = H/d_p$	$h = H\ (\mu m)/d_p\ (\mu m)$ $= \dfrac{10^3}{5.54} \cdot \dfrac{L\ (mm)}{d_p\ (\mu m)} \cdot \left[\dfrac{tr}{w_{½ht}}\right]^2$
linear velocity	$u = L/t_o$	$u\,(mm/s) = L\,(mm)/t_o\ (s)$
reduced velocity	$\nu = \dfrac{u\,d_p}{D_m} = \dfrac{L d_p}{t_o D_m} = \dfrac{l\,d_p^2}{t_o D_m}$	$\nu = \dfrac{L\,(mm)\ d_p(\mu m) \times 10^{-9}}{t_o\ (s)\ D_m\ (m^2/s)}$
reduced length	$l = L/d_p = Nh$	$l = 1000\ L\,(mm)/d_p\ (\mu m)$
permeability	$K = \dfrac{u\eta L}{P} = \dfrac{\eta L^2}{P t_o}$	$K^\circ(m^2) = \dfrac{\eta\,(Ns/m^2)\ L^2\,(mm) \times 10^{-9}}{P\,(bar)\ t_o\ (s)}$
resistance	$\phi' = d_p^2/K_o = \dfrac{P t_o d_p^2}{\eta L^2}$ $= \dfrac{P t_o}{\eta l^2} = P/(l^2 \eta u)$	$\phi' = \dfrac{0.1\ P\,(bar)\ t_o\ (s)\ d_p^2\ (\mu m)}{\eta\ (Ns/m^2)\ L^2\ (mm)}$ [η for water at 25°C = 1cP = 10^{-3} Ns/m^2]
separation impedance	$E = h^2 \phi' = H^2/K^\circ$	$E = 0.036 \left[\dfrac{t_r}{w_{½ht}}\right]^4 \cdot \dfrac{P\,(bar)\ t_o\ (s)}{\eta\ (Ns/m^2)}$
column tube volume	$\chi = (\pi/4)\ d_c^2\ L$	$\chi\,(ml) = 0.785\ d_c^2\ (mm)\ L\,(m)$

(Table of Symbols, section 11.6, page 228)

LC in practice P A Bristow **hetp**, Macclesfield UK ISBN 0 9504833 0 3